PHARMACY TECHNICIAN REVIEW and TEST PREPARATION

MARVIN M. STOOGENKE, *B.S., R.Ph.*

BRADY
PRENTICE HALL
Upper Saddle River, New Jersey 07458

Library of Congress Cataloging-in-Publication Data

Stoogenke, Marvin M.
    Pharmacy technician review and test preparations / Marvin M. Stoogenke.
      p. cm.
    ISBN 0-8359-5328-9 (p)
    1. Pharmacy technicians—Examinations, questions, etc.   I. Title.
[DNLM: 1. Pharmacists' Aides examination questions. 2. Pharmacy examination questions. QV 18.2S882p   1999]
RS122.95.S763   1999
615'.1'076—dc21
DNLM/DLC
for Library of Congress   98-11720
                                                                                   CIP

Publisher: Susan Katz
Acquisitions Editor: Barbara Krawiec
Editorial Assistant: Stephanie Carmangian
Marketing Manager: Tiffany Price
Marketing Coordinator: Cindy Frederick
Director of Production and Manufacturing: Bruce Johnson
Managing Production Editor: Patrick Walsh
Senior Production Manager: Ilene Sanford
Production Editor: Larry Hayden IV
Creative Director: Marianne Frasco
Cover design: Miguel Ortiz
Composition: BookMasters, Inc.
Presswork and Binding: Banta Press, Harrisonburg, VA

Reviewers:

Susan Balducci, BSN, MS
Program Head, Western Wisconsin Technical College
La Crosse, Wisconsin

Lt. Col. Stephen Wortham, NREMT-P, PI
Yellow Ambulance Service–Education Specialist
Louisville, Kentucky

© 1999 by Prentice-Hall, Inc.
A Pearson Education Company
Upper Saddle River, NJ 07458

All rights reserved. No part of this book may be reproduced, in any form or by any means, without permission in writing from the publisher.

Printed in the United States of America

10 9 8 7 6 5 4

ISBN: 0-8359-5328-9

Prentice-Hall International (UK) Limited, London
Prentice-Hall of Australia Pty. Limited, Sydney
Prentice-Hall Canada Inc., Toronto
Prentice-Hall Hispanoamericana, S.A., Mexico
Prentice-Hall of India Private Limited, New Delhi
Prentice-Hall of Japan, Inc., Tokyo
Pearson Education Asia Pte. Ltd., Singapore
Editora Prentice-Hall do Brasil, Ltda., Rio de Janeiro

Contents

Preface v

Introduction 1

QUESTIONS 5

I Assisting the Pharmacist in Serving Patients 5
II Medication Distribution and Inventory Control Systems 148
III Operations 153

ANSWERS 172

I Assisting the Pharmacist in Serving Patients 172
II Medication Distribution and Inventory Control Systems 201
III Operations 202

Disclaimer

This review text is specifically designed to present information to guide the reader in preparing for pharmacy technician certification examinations. The author and publisher have made a conscientious effort to assure accuracy, completeness, and compatibility with the standards generally accepted for information presented at the time of publication. Nevertheless, as drug information becomes available as a result of research and experience, changes in the use of drugs and pharmacy practice become necessary. Also, certification examinations may vary from offering to offering. The reader is advised that the author and publisher cannot be responsible for any errors or omissions arising from new information, interpretations, or changes in the scope of the examination.

Preface

Pharmacy Technician Review and Test Preparation is a basic review guide to help you structure your study for the national certification examination and to assist you with classroom work or on-the-job training programs. This review book contains an abundance of questions to prepare you for taking tests. The questions are constructed to assist you in studying for each section of the certification examination for pharmacy technicians.

The National Pharmacy Technician Certification Examination is used to test the qualifications of individuals working in a pharmacy setting who work under the supervision of a licensed pharmacist. The role, duties, and responsibilities of the pharmacy technician are supportive. The pharmacy technician assists the pharmacist in delivering services. Hence, the examination focuses predominantly on "Assisting the Pharmacist in Serving Patients." This area of the examination is worth 50% of the total score. It focuses on traditional distributive pharmacy activities which involve the processing of drug orders. The pharmacy technician's job includes receiving the drug request by prescription (traditionally used in the outpatient setting) or medication order (typically found in the inpatient or institutional setting). Competence is measured by the ability to process new orders or refills received in various formats (hardcopy or electronic) as well as thoroughness in assessing authenticity and patient information regarding allergies, sensitivities, previous adverse reactions, and illness.

Questions test the candidate's knowledge of products, calculations, conversions, and terminology. (NOTE: drug products are indicated by brand or trade names in capital letters and generic names in lower case letters). Various specifics within pharmacy practice are covered and include compounding, intravenous admixture preparation, packaging, labeling, and various types of record keeping.

The second area of the examination deals with "Medication Distribution and Inventory Control Systems." This section contributes 35% to the overall test. Activities in this section include the handling of drugs to assure their availability. Pharmacy technicians are expected to understand the purchasing process, inventory control, and stocking in accordance with approved job-centered policies and procedures. The candidate should be aware of the way orders to vendors are placed, the activities initiated when goods are received, and the controls which assure orders can be filled using available stock without exceeding the pharmacy's budget. The pharmacy technician should note that stock in many cases includes more than drugs. Stock may also include durable medical equipment inventory, devices, and supplies.

The third and final area of concentration, "Operations," includes administrative functions pertinent to pharmacy operations. These include human resources, facilities and equipment maintenance, communications, policies and procedures, compliance with regulations and professional standards, and the use of computer-systems.

Studying this review text in conjunction with the author's textbook, *The Pharmacy Technician,* a drug handbook and a medical dictionary should provide abundant information covering each area of the examination. Preparation, as well as a good night's rest, is essential to successful completion of the examination.

Acknowledgments

I would like to express my appreciation and thanks to my family for their continuing support of my writing activities. Thanks go to my sons Scott, Jason, and Saul, who never cease to lend an ear and a word. Special thanks go to my wife, Judy, and Saul, whose seemingly tireless typing of the manuscript assured its timeliness.

Judy, your perseverance and energy keep me going regardless of my thoughts on the matter. You give me purpose and you are my purpose. Thank you. I love you. MWA

Additionally, I extend my gratitude to the Pharmacy Technician Certification Board visionaries who were able to see the role pharmacy technicians would play in an evolutionary health-care delivery system, and who have pioneered to define the role and importance qualified pharmacy technicians have in the health of society.

Introduction

Health care is dynamic. Pharmacy as part of this dynamic environment is in constant evolution. Changes mean new roles for the current cast and openings for new cast members. Although pharmacy technicians have been practicing for some time, until now their role was rudimentary. The traditional pharmacy technician supported distributive services under the supervision of a pharmacist by performing simple tasks. These included filling the prescription with the appropriate drug, packaging the medication in the proper container, and affixing the label containing the directions for use as intended by the prescriber. Pharmacy practice, regardless of the setting, was basic and the pharmacy technician's role was, likewise, basic. The fundamental tenet guiding the pharmacy technician has been to provide the right drug to the right patient at the right time.

As health-care delivery changes to accommodate the needs and wants of patients, legislative mandates, technology, and much more, the services associated with pharmacy practice also change. Pharmacists who practice only distributive pharmacy are recognizing that they must expand their roles to include providing information to patients and to other health-care professionals. In addition, pharmacists have a more significant role in other non-distributive pharmacy practices such as cost containment, formulary services, and third-party payers. The pharmacy technician closes the potential gap in traditional pharmacy delivery that these major changes in health care are creating. However, in order to continue to deliver the traditional services of pharmacy without interruption in patient care, pharmacy technicians must be qualified by having the proper knowledge, appropriate skills, and competence to provide these services.

In order to assure the smooth operation of distributive pharmacy, the Pharmacy Technician Certification Board certification program was developed to assist in defining the role of pharmacy technicians. Certification is a significant career progression that establishes uniformity of knowledge, consistency in applying that knowledge, and the basis for a standard of practice. The National Pharmacy Technician Certification Examination is a formal vehicle by which pharmacy technicians are determined to be qualified to perform the tasks associated with distributive pharmacy practice.

This book is the first edition of the *Pharmacy Technician Review and Test Preparation*. The author designed this text to assist the reader in studying for and successfully completing the National Pharmacy Technician Certification Examination. In addition, this review guide is a valuable tool for students in on-the-job training programs or in formalized pharmacy technician training programs. It can be used in conjunction with a pharmacy textbook such as *The Pharmacy Technician,* a drug handbook such as the *Drug Information Handbook,* and a medical dictionary. The questions are comprehensive, easily read and aligned with the national examination. As an enhancement to the learning experience, many questions were derived from actual pharmacy events.

The qualified pharmacy technician applies the principles of accurate and safe pharmacy practice. This is accomplished through knowledge of drug names and uses, medical and pharmacy terminology and abbreviations, calculations, laws regulating practice, and common disease states, symptoms, and the medications used to treat them. The pharmacy technician's responsibilities and activities are governed by standards of ethics. These standards direct him/her in the selection of the appropriate drug, correct strength and form, and the preparation of understandable label directions for the patient or caregiver.

This review book tests the extent of knowledge required to do the job while assuring that the patient's safety and well-being are primary. The author has tried to correlate principles derived from pharmacy ethics with the information required for the certification examination. Certification is a means of establishing credibility. Certification is the first step.

Author's Personal Note

As a pharmacist, I have been associated with pharmacy technicians for many years. As an author, I have provided a basis for anyone interested in becoming a pharmacy technician to learn a uniform body of knowledge in a practical manner. My experiences have been gratifying. Pharmacy technicians have often impressed me with their know-how and enthusiasm.

The future for pharmacy technicians looks bright and full of opportunities. The changes continuously occurring in health-care delivery will benefit the pharmacy technician's career prospects. Only now are we noticing the importance of the technician on a global scale. The pharmacy technician will play a key role in keeping distributive pharmacy

Introduction

alive and pharmacy costs in line with the mandate for prudent health-care economics.

You are at an ideal point from which to chart your course. You need only remember that there are four components to your successful future.

KNOWLEDGE is the first component. Whether you learn the basic tools of pharmacy in the classroom, at the job-site, or from books on your own, you have the key to opening a door of exciting challenges.

Your EXPERIENCE will be your second component. Real world experience provides the ultimate practical learning environment. Some call it "street-savvy," but this will enable you to deal with real people in a real way. As you become an effective participant in health-care delivery, you will serve well both the patient individually and the society as a whole.

The third component is SELF-PACING. We *never dispense guesswork*. Pace yourself until you understand what must be understood. Proceed only when you feel comfortable. Your grasp of the knowledge will assure your success on the certification exam, in the workplace, and within yourself. The extra time is a small price to invest for a fulfilling future.

Finally, there is the component of SELF-PRESENTATION. This is an attitude which enables you to walk proudly among other professionals. Think professionally, act professionally, speak professionally, and present yourself professionally.

Good luck in your career.

Marvin M. Stoogenke, R.Ph.

Questions

I. Assisting the Pharmacist in Serving Patients

Drugs

> **Brand & Generic Names**
>
> **Primary Uses**
>
> **Usual Dosing**
>
> **Common Side Effects**
>
> **Notable Interactions**
>
> **Special Information**
>
> **Classifications & Groupings**
>
> 1. Ergotamine is to CAFERGOT as sucralfate is to ____?
>   1. BANTHINE
>   2. QUARZAN
>   3. CARAFATE
>   4. ROBINAL
> 2. Which of the following answers best describes a primary indicated use for enalapril?
>   1. atherosclerosis
>   2. arrhythmia

3. hypertension
4. bradycardia

3. What is the brand name for ketorolac?
   1. TORADOL
   2. TOBREX
   3. TOFRANIL
   4. KYTRIL

4. Which of the following is the generic name for PREMARIN?
   1. tamoxifen
   2. norethindrone
   3. ethinyl estradiol
   4. conjugated estrogens

5. ORTHO-NOVUM is to NORINYL as diltiazem is to ____?
   1. phenobarbital
   2. verapamil
   3. hydantoin
   4. ranitidine

6. Which of the following answers best describes the most common indication for fluoxetine?
   1. bipolar affective disorders (BAD)
   2. seasonal affective disorders (SAD)
   3. hypertension
   4. major depression

7. Which is the generic name for HUMULIN?
   1. insulin
   2. glipizide
   3. glyburide
   4. chlorpropamide

8. How would you classify or group sertraline?
   1. NSAID
   2. HMG-CoA
   3. TCA
   4. SSRI

9. Which is a trade name for penicillin VK?
   1. VERELAN
   2. PENTIDS
   3. V-CILLIN K
   4. VISKEN

10. How would you classify or group levothyroxin?
    1. hypoglycemic agent
    2. thyroid hormone
    3. corticosteroid
    4. contraceptive agent

11. Which of the following is the generic name for COUMADIN?
    1. warfarin
    2. hydrocortisone
    3. triamcinolone
    4. cortisone

12. Which of the following is a side effect reported with sertraline?
    1. loss of muscle coordination
    2. chest pain
    3. dry mouth
    4. rash
13. Which of the following is the trade name for terfenadine?
    1. SELDANE
    2. CLARITIN
    3. HISMANAL
    4. BENADRYL
14. Which of the following answers best describes the usual oral dosage for loratidine?
    1. 10 mg. daily
    2. 10 mg. BID
    3. 20 mg. daily
    4. titrate dose as needed
15. Which is an appropriate auxiliary label to use with nabumetone?
    1. AVOID ALCOHOL
    2. AVOID ASPIRIN
    3. TAKE WITH FOOD OR MILK
    4. all of the above
16. Which of the following answers best describes the primary indicated use for ethinyl estradiol/levonorgestrel?
    1. inactive thyroid
    2. hypoglycemia
    3. prostatic cancer
    4. prevention of pregnancy
17. How would you classify or group penicillin VK?
    1. antifungal
    2. anti-infective
    3. antiprotozoal
    4. antiulcer
18. Which is an appropriate auxiliary label to use with verapamil?
    1. TAKE WITH FOOD (sustained release form)
    2. DO NOT CRUSH (sustained release form)
    3. STORE IN GLASS (sustained release form)
    4. answers 1 and 2
19. Which of the following is the generic name for MYOCHRISINE?
    1. gold sodium thiomalate
    2. gonadorelin
    3. goserelin
    4. granisetron
20. Which of the following best describes a unique characteristic for colchicine?
    1. causes localized rash
    2. overdose causes hair loss
    3. causes heartburn
    4. impairs the absorption of vitamin B-12

21. Is nizatidine available as an OTC?
    1. Yes
    2. No
    3. Pending
    4. Under FDA review
22. What information should you give the patient taking simvastatin?
    1. Use in conjunction with dietary therapy.
    2. Double missed doses.
    3. Report muscle pain, fever, weakness.
    4. answers 1 and 3
23. What is the generic name for ADALAT?
    1. nisoldipine
    2. nizatidine
    3. acebutolol
    4. nifedipine
24. Which of the following drugs is a selective serotonin reuptake inhibitor (SSRI)?
    1. diazepam
    2. chlorpromazine
    3. fluoxetine
    4. haloperidol
25. Which of the following answers best describes the primary indicated use of verapamil?
    1. edema
    2. angina, hypertension
    3. TIA
    4. MI, CHF
26. Which is a trade name for triamterene/hydrochlorothiazide?
    1. MAXZIDE
    2. DIUPRES
    3. DYRENIUM
    4. MATULANE
27. Clonidine is to ____, as quinidine is to QUINORA.
    1. TENEX
    2. CATAPRES
    3. ALDOMET
    4. WYTENSIN
28. Coadministration of simvastatin and which of the following may produce an unwanted result?
    1. herbal medications
    2. gemfibrozil
    3. loperamide
    4. diphenoxylate
29. Which of the following answers best describes the primary indications for ranitidine?
    1. treat angina, hypertension
    2. treat hypercholesterolemia

Drugs

3. treat depression, obsessive-compulsive disorder
4. treat duodenal ulcer, gastroesophageal reflux, and gastric hypersecretory conditions

30. Which of the following is the generic name for LASIX?
    1. bumetanide
    2. amiloride
    3. spironolactone
    4. furosemide

31. Which of the following answers best describes the primary indicated use for ethinyl estradiol/norethindrone?
    1. prevention of pregnancy
    2. supplement to breast cancer treatment
    3. prostatic carinoma
    4. inactive thyroid

32. Which best describes the usual adult dose for clonazepam?
    1. up to 5 cc./d
    2. not to exceed 10 units/d
    3. 15 mcg./Kg. body weight
    4. maximum 20 mg./d

33. What is the brand name for pravastatin?
    1. PROCARDIA
    2. PRECOSE
    3. PRONESTYL
    4. PRAVACHOL

34. Which of the following is the generic name for PLAQUENIL?
    1. hydroxocobalamin
    2. plicamycin
    3. hydroxyprogesterone
    4. hydroxychloroquine

35. Which answer best describes the usual adult dose for penicillin VK?
    1. up to 2 grams per Kg. per day
    2. up to 2 grams per day
    3. 250 mcg. q6h around the clock
    4. answers 1 and 2

36. Which of the following is a common side effect reported with clonazepam?
    1. drowsiness
    2. tachycardia
    3. bradycardia
    4. rash

37. Which of the following is the generic name for PROZAC?
    1. fluconazole
    2. fluoxetine
    3. fluphenazine
    4. flurazepam

38. Which of the following is the generic name for TAXOL?
    1. pamidronate
    2. paclitaxel

3. pancrelipase
4. tamoxifen

39. Which of the following answers best describes a unique association with nifedipine?
    1. Take medication with plenty of fluid.
    2. Do not chew or break sustained-release dosage form.
    3. Store in glass.
    4. Limit time spent in sunlight.

40. Coadministration of glipizide and which of the following may produce an unwanted result?
    1. TCN
    2. PCN
    3. NSAIDS
    4. DPH

41. Which of the following is the trade name for glipizide?
    1. DIABETA
    2. GLUCOTROL
    3. ORINASE
    4. TOLINASE

42. Which of the following is a side effect reported with penicillin VK?
    1. bruising
    2. thirst
    3. diarrhea
    4. muscle cramps

43. How would you classify or group simvastatin?
    1. antiemetic agent
    2. antihyperlipidemic agent
    3. antidiarrheal agent
    4. anti-inflammatory agent

44. Using the key phrase, "gastric acid secretion" as the basis for your decision, which of the following does not belong?
    1. diphenhydramine
    2. omeprazole
    3. nizatidine
    4. cimetidine

45. Which of the following is the generic name for DIFLUCAN?
    1. ketoconazole
    2. fluconazole
    3. itraconazole
    4. miconazole

46. "Azepam" in the name of a drug best describes which of the following types of drugs?
    1. beta lactams
    2. benzodiazepines
    3. steroids
    4. tricylic antidepressants

47. Which of the following answers best describes the most common indication(s) for omeprazole?
    1. gastroenteritis
    2. angina, hypertension
    3. Crohn's disease
    4. erosive esophagitis and gastroesophageal reflux disease
48. Which of the following answers best describes the primary indicated use for sertraline?
    1. cardiovascular disease
    2. major depression
    3. peptic ulcer disease
    4. chronic obstructive pulmonary disease
49. Which of the following answers best describes the usual dosage for paclitaxel?
    1. 500 mg./Kg. every 8 hours
    2. 250 mg. every 4 hours
    3. 135–175 mg./sq. meter over 3 hours every 3 weeks
    4. 1 Gm. bolus
50. Which is an appropriate auxiliary label to use with penicillin VK?
    1. TAKE WITH ORANGE JUICE
    2. FINISH ALL THIS MEDICATION UNLESS OTHERWISE DIRECTED BY PRESCRIBER
    3. MAY CAUSE DROWSINESS
    4. MAY CAUSE DISCOLORATION OF URINE OR FECES
51. Which of the following is a side effect reported with hydroxychloroquine?
    1. digital numbness
    2. joint pains
    3. gastrointestinal disturbance
    4. sweats
52. Which is an appropriate auxiliary label to use with methotrexate?
    1. TAKE ON AN EMPTY STOMACH
    2. TAKE WITH FOOD
    3. DO NOT CRUSH
    4. AVOID SMOKING WHILE TAKING THIS MEDICATION
53. Which of the following is a side effect reported with simvastatin?
    1. abdominal cramps
    2. constipation
    3. headache
    4. all of the above
54. Amoxicillin may diminish the action of which group of drugs?
    1. antidepressants
    2. steroids
    3. oral contraceptives
    4. antihistamines
55. Which of the following is the generic name for ANAPROX?
    1. nalaxone
    2. amrinone

3. naproxen
4. nandrolone

56. What information should you give the patient taking ethinyl estradiol/levonorgestrel?
    1. Reduce dietary fiber while taking this medication.
    2. Discontinue if you are pregnant or intend to become pregnant.
    3. Double missed doses.
    4. Skip a dose if experiencing vision disturbance.

57. What information should you give the patient taking penicillin VK?
    1. Complete the full course of therapy.
    2. Do not go to work while on medication.
    3. Do not take OTC medications while taking this medication.
    4. Double up taking a multivitamin.

58. Using the key word, "lipids" as the basis for your decision, which of the following does not belong?
    1. lovastatin
    2. gemfibrozil
    3. nicotinic acid
    4. nystatin

59. Which is an appropriate auxiliary label to use with warfarin?
    1. MAY CAUSE DISCOLORATION OF THE URINE OR FECES
    2. DO NOT TAKE DAIRY PRODUCTS WHILE TAKING THIS MEDICATION
    3. AVOID PROLONGED EXPOSURE TO SUNLIGHT
    4. TAKE WITH FOOD OR MILK

60. What is the trade name for lisinopril?
    1. ZESTRIL
    2. DIURIL
    3. PRINVIL
    4. answers 1 and 3

61. What is the generic name for the drug CECLOR?
    1. cephalexin
    2. cephradine
    3. cefuroxime
    4. cefaclor

62. Which of the following is the trade name for cyclophosphamide?
    1. CYLERT
    2. CYTOXAN
    3. CYTADREN
    4. CYTOGAM

63. Which of the following is the generic name for PEPCID?
    1. ranitidine
    2. cimetidine
    3. famotidine
    4. nizatidine

64. What information should you give the patient or caregiver about taking terfenadine?
    1. Monitor the drying effect on an asthmatic patient.
    2. Medication may cause drowsiness.

Drugs

3. Drink plenty of fluid.
4. all of the above

65. Coadministration of penicillin VK and which of the following may produce an unwanted result?
    1. atenolol
    2. carisoprodol
    3. tetracycline
    4. sertraline

66. Cimetidine is contraindicated in patients treated with which of the following?
    1. theophylline
    2. warfarin
    3. zolpidem
    4. answers 1 and 2

67. Which of the following answers best describes the most common indication(s) for lovastatin?
    1. angina, hypertension
    2. hypercholesterolemia
    3. hypercalcemia
    4. hypernatremia

68. Which of the following answers best describes the usual adult oral dosage for phenytoin?
    1. maintenance dose: one capsule TID or QID
    2. maximum dose: two capsules TID
    3. once-a-day dose: 300 mg. of the extended release form
    4. all of the above

69. Which of the following is a common side effect reported with diclofenac use?
    1. headache
    2. gastrointestinal disturbance
    3. lightheadedness
    4. loss of hair

70. Which of the following is the generic name for TRIPHASIL 28?
    1. ethinyl estradiol
    2. ethinyl estradiol/levonorgestrel
    3. conjugated estrogens
    4. ethinyl estradiol/norethindrone

71. Coadministration of warfarin and which of the following may result in a potentially serious or even life-threatening drug interaction?
    1. barbiturates
    2. H2 antagonists
    3. sulfa antibiotics
    4. all of the above

72. What information should you give the patient taking nifedipine?
    1. Do not take any herbal medications while on this medication.
    2. Rise slowly from a lying or sitting position.
    3. Double up on missed doses.
    4. Adjust doses as needed.

73. Using the key phrase, "gram positive" as the basis for your decision, which of the following does not belong?
    1. cephalosporins
    2. penicillins
    3. erythromycin
    4. capreomycin
74. Coadministration of furosemide and which of the following may produce an unwanted result?
    1. steroids
    2. analgesics
    3. hypoglycemic agents
    4. histamine H2 antagonists
75. Using the key phrase, "seasonal allergy" as the basis for your decision, which of the following does not belong?
    1. diphenhydramine
    2. diphenylhydantoin
    3. loratidine
    4. terfenadine
76. Which of the following is the trade name for loratidine?
    1. CLAFORAN
    2. CLEOCIN
    3. LOPID
    4. CLARITIN
77. Which of the following is the generic name for VASOTEC?
    1. enalapril
    2. verapamil
    3. lisinopril
    4. amlodipine
78. Which of the following is the trade name for prazosin?
    1. MINOCIN
    2. PRILOSEC
    3. MINIPRESS
    4. MONOPRIL
79. Which of the following answers best describes the primary indicated use for insulin?
    1. diabetes insipidus
    2. lipoatrophic diabetes
    3. insulin-dependent diabetes mellitus
    4. noninsulin-dependent diabetes mellitus
80. Which is a brand name for penicillin VK?
    1. PENTAM
    2. PEN VEE K
    3. PENTIDS
    4. PECID
81. Which of the following is the generic name for RELAFEN?
    1. ranitidine
    2. nadolol
    3. rifampin
    4. nabumetone

Drugs

82. Which of the following drug classes or groups is represented by nabumetone?
    1. analgesics (opiate)
    2. anti-inflammatory agents (NSAID)
    3. psychotherapeutic agent (tranquilizer)
    4. cardiovascular agent (ACE inhibitor)

83. What is the trade name for nifedipine?
    1. NARDIL
    2. PRONESTYL
    3. ADALAT
    4. PROCAN

84. Which best describes the usual adult dose for colchicine?
    1. 8 mg. BID
    2. maximum of 8 mg. per course of therapy
    3. 2 mg. TID
    4. up to 4 mg. per day separated by 3 day intervals

85. Which of the following is the generic name for ZOFRAN?
    1. zolpidem
    2. ofloxacin
    3. omeprazole
    4. ondansetron

86. Which of the following is a common side effect reported with ethinyl estradiol/norethindrone use?
    1. peripheral edema
    2. tinnitus
    3. metallic taste
    4. hallucinations

87. Which of the following is the generic name for DYAZIDE?
    1. triamterene
    2. chlorothiazide
    3. triamterene/hydrochlorothiazide
    4. bumetanide

88. Which is an appropriate auxiliary label to use with loratidine?
    1. TAKE WITH FOOD
    2. FINISH ALL THIS MEDICATION UNLESS OTHERWISE DIRECTED BY PRESCRIBER
    3. MAY CAUSE DISCOLORATION OF URINE OR FECES
    4. TAKE ON AN EMPTY STOMACH

89. Ciprofloxacin works best when avoiding which of the following?
    1. antacids containing aluminum, magnesium, or calcium
    2. products containing zinc or iron
    3. caffeine
    4. all of the above

90. How would you classify or group terfenadine?
    1. analgesic
    2. antihistamine
    3. diuretic
    4. hormone

91. Using the key word, "hyperglycemic" as the basis for your decision, which of the following does not belong?
    1. thirst
    2. visual disturbance
    3. excessive urination
    4. none of the above
92. What is the tradename for hydroxychloroquine?
    1. HYDREA
    2. PLACIDYL
    3. PLAQUENIL
    4. PLENDIL
93. Which of the following is the generic name for VOLTAREN?
    1. dicloxacillin
    2. dicyclomine
    3. verapamil
    4. diclofenac
94. How would you classify or group penicillamine?
    1. antibiotic
    2. antiviral
    3. antifungal agent
    4. chelating agent
95. Which of the following answers best describes the primary indicated use for ketorolac?
    1. cardiac arrhythmia
    2. indigestion
    3. GERD
    4. short term pain management
96. What is the generic name for PROCARDIA?
    1. nizatidine
    2. nisoldipine
    3. nifedipine
    4. procarbazine
97. Which of the following best describes the primary indicated use for simvastatin?
    1. hyperkalemia
    2. hyponatremia
    3. hypercholesterolemia
    4. hypoglycemia
98. Which of the following drug classes or groups is represented by medroxyprogesterone?
    1. hormones (progestogens)
    2. hormones (adrenal corticosteroids)
    3. hormones (anabolic steroid)
    4. hormones (estrogens)
99. What is the generic name for BIAXIN?
    1. erythromycin
    2. clarithromycin
    3. streptomycin
    4. vancomycin

100. Which of the following answers best describes the most common indication for glyburide?
    1. hypoglycemia
    2. hyperglycemia
    3. hypercholesterolemia
    4. hypertension
101. Which of the following answers best describes the primary indicated use for penicillin VK?
    1. malaria
    2. fungal infections
    3. HIV infection
    4. bacterial infection
102. Which answer best describes the usual adult dose for nabumetone?
    1. up to 2 Gm./day
    2. up to 2 mg./day
    3. up to 2 mcg./day
    4. 2 mg. BID
103. What information should you emphasize to the patient taking glyburide?
    1. the importance of good foot care
    2. the importance of immediate care for cuts and bruises
    3. the importance of not skipping meals
    4. all of the above
104. What do amitriptyline, imipramine, nortriptyline, and clomipramine have in common?
    1. calcium channel blocker
    2. SSR inhibitors
    3. MAO inhibitors
    4. TCA's
105. Which of the following is a side effect reported with terfenadine?
    1. glossitis
    2. rapid heart rate
    3. memory loss
    4. dry mouth
106. What is the trade or brand name for the generic grug captopril?
    1. VASOTEC
    2. MONOPRIL
    3. LOTENSIN
    4. CAPOTEN
107. Coadministration of omeprazole and which of the following may result in a potentially serious drug interaction?
    1. NSAIDs
    2. antihistamines
    3. hydantoins
    4. fluoroquinolones
108. Using the key word, "insulin" as the basis for your decision, which of the following does not belong?
    1. glyburide
    2. glipizide

3. furosemide
4. tolazamide

109. Which answer best describes the usual adult dose for paroxetine?
    1. 50 mg. qhs
    2. up to 50 mcg. per day
    3. 25 mcg. BID
    4. maximum 50 mg. per day

110. Which of the following is the trade name for furosemide?
    1. LASIX
    2. LANOXIN
    3. LARODOPA
    4. FORTAZ

111. What information should you give the patient taking triamterene/hydrochlorothiazide?
    1. Supplement your diet with a banana per day.
    2. Take medication early enough during the day to prevent nocturia.
    3. Double your vitamin intake per day.
    4. Double up on missed doses.

112. Which of the following is the generic name for KLONOPIN?
    1. clonazepam
    2. clorazepate
    3. clonidine
    4. clozapine

113. How would you classify or group hydroxychloroquine?
    1. antimalarial agent
    2. cardiovascular agent
    3. musculoskeletal agent
    4. psychotherapeutic agent

114. What is the trade or brand name for the generic drug paclitaxel?
    1. PAXIL
    2. TAXOL
    3. TALWIN
    4. PARLODEL

115. What is the generic name for the drug CAPOTEN?
    1. ramipril
    2. captopril
    3. lisinopril
    4. enalapril

116. Which of the following is a side effect reported with penicillamine?
    1. fever
    2. joint pain
    3. lessened ability to taste
    4. all of the above

117. What is the trade or brand name for the generic drug medroxyprogesterone?
    1. OGEN
    2. MICRONOR
    3. PROVERA
    4. OVRETTE

118. Coadministration of terfenadine and which of the following may result in a serious or even life-threatening drug interaction?
    1. beta-lactam antibiotics
    2. macrolide antibiotics
    3. antifungal drugs
    4. answers 2 and 3
119. What information should you give the patient taking diclofenac?
    1. Do not crush tablets.
    2. Take with food or milk.
    3. Report signs of blood in the stools.
    4. all of the above
120. Which answer best describes the usual adult dose for ibuprofen?
    1. 800 mg. daily
    2. two tablets QID
    3. 400 mg. qhs
    4. maximum 3.2 Gm. per day
121. What is the generic name for the drug PROVERA?
    1. estropipate
    2. norgestrel
    3. norethindrone
    4. medroxyprogesterone
122. Which of the following is a common side effect reported with conjugated estrogens?
    1. breast enlargement and tenderness
    2. peripheral edema
    3. bloating
    4. all of the above
123. Which is an appropriate auxiliary label to use with medroxyprogesterone?
    1. DO NOT CRUSH
    2. AVOID PROLONGED EXPOSURE TO SUNLIGHT WHILE TAKING THIS MEDICATION
    3. TAKE ON AN EMPTY STOMACH
    4. MAY CAUSE DROWSINESS
124. Coadministration of naproxen and which of the following may result in a potentially serious or even life-threatening drug interaction?
    1. glyburide
    2. methotrexate
    3. loperamide
    4. acetaminophen
125. What is the brand name for propoxyphene napsylate/acetaminophen?
    1. DARVOCET-N
    2. DARVON-N
    3. DARVON COMPOUND
    4. DECADRON
126. Which of the following is the generic name for CUPRIMINE?
    1. penicillin
    2. carbamazepine

3. penicillamine
4. pentamidine

127. Which of the following answers best describes the primary indicated use for nabumetone?
    1. manage pain associated with angina
    2. manage pain associated with dysmenorrhea
    3. manage pain and inflammation associated with osteoarthritis and rheumatoid arthritis
    4. manage mild to moderate hypertension

128. How would you classify or group ketorolac?
    1. G-CSF
    2. ACE inhibitor
    3. NSAID
    4. opiate analgesic

129. Which of the following answers best describes a unique association with glipizide?
    1. increased risk of cardiovascular mortality
    2. increased risk of impotence
    3. increased risk of ototoxicity
    4. increased risk of irreversible visual dysfunction

130. How would you classify or group triamterene/hydrochlorothiazide?
    1. analgesic
    2. hypoglycemic
    3. dermatologic
    4. diuretic

131. Which of the following is the trade name for fluconazole?
    1. NIZORAL
    2. SPORANOX
    3. LOTRIMIN
    4. DIFLUCAN

132. Which of the following is the generic name for PAXIL?
    1. paroxetine
    2. paclitaxel
    3. penicillamine
    4. pamidronate

133. What is the generic name for the drug CIPRO?
    1. ofloxacin
    2. ciprofloxacin
    3. lomefloxacin
    4. norfloxacin

134. What information should you give the patient taking penicillamine?
    1. Take medication at least 1 hour before meals.
    2. Drink fluids liberally.
    3. May lose taste sensation.
    4. All of the above.

135. Which of the following answers best describes the primary indicated use for terfenadine?
    1. edema
    2. symptoms of seasonal allergy

3. mild to moderate pain
4. ERT

136. Which is a brand name for triamterene/hydrochlorothiazide?
    1. DYRENIUM
    2. DYAZIDE
    3. HYDRODIURIL
    4. none of the above

137. BENADRYL is to BENTYL as antihistamine is to ____?
    1. antidepressant
    2. antibiotic
    3. antispasmodic
    4. antihypertensive

138. Which of the following answers best describes the most common indication(s) for diltiazem?
    1. edema
    2. angina, hypertension
    3. congestive heart failure, MI
    4. atherosclerosis

139. Which of the following answers best describes the usual oral dosage for ciprofloxacin?
    1. Dose and frequency depend on the type and severity of the infection.
    2. All doses have a schedule of q12h.
    3. Daily doses range from 500–1500mg.
    4. all of the above

140. Which of the following is a common side effect reported with cyclophosphamide use?
    1. headache
    2. lightheadedness
    3. dizziness
    4. hair loss

141. Calcium channel blockers are contraindicated in patients treated with which of the following?
    1. quinine medications
    2. quinapril
    3. quinidine products
    4. quinestrol

142. Which of the following is a side effect reported with triamterene/hydrochlorothiazide?
    1. bruising
    2. loss of appetite
    3. hair loss
    4. visual disturbances

143. What information should you give the patient taking medroxyprogesterone?
    1. Adjust dosage as necessary.
    2. Take medication exactly as directed.
    3. Take medication with orange juice.
    4. Double missed doses.

144. Which of the following answers best describes the primary indicated use for naproxen?
    1. RA
    2. dysmenorrhea
    3. inflammatory disease
    4. all of the above

145. Coadministration of penicillamine and which of the following may produce an unwanted result?
    1. diphenhydramine
    2. iron salts
    3. cimetidine
    4. acidic juices

146. What is the brand name for ibuprofen?
    1. ORUDIS
    2. RELAFEN
    3. MOTRIN
    4. DAYPRO

147. What do phenelzine and tranylcypromine have in common?
    1. MAO inhibitors
    2. SSR inhibitors
    3. NSAIDs
    4. TCAs

148. What information should you give the patient taking verapamil?
    1. Limit caffeine intake.
    2. Maintain a low-sodium diet.
    3. Use available techniques to deal with stress.
    4. all of the above

149. What is the generic name for the drug CLARITIN?
    1. terbutaline
    2. diphenhydramine
    3. clemastine
    4. loratidine

150. What classification is given to ceftibuten?
    1. first-generation cephalosporin
    2. second-generation cephalosporin
    3. third-generation cephalosporin
    4. carbacepham derivative

151. Elavil is to amitriptyline as ____ is to MELLARIL.
    1. theophylline
    2. thioridazine
    3. thiothixene
    4. thiopental

152. What is the usual adult oral dose for diltiazem?
    1. capsules, up to 500 mg. per day
    2. tablets, up to 600 mg. per day
    3. tablets or capsules, 180 mg. per day
    4. Optimum maintenance dose does not exceed 360 mg./d.

Drugs

153. Beta blockers are contraindicated in patients treated with which of the following?
    1. clonidine
    2. cladribine
    3. clemastine
    4. none of the above
154. Using the key word, "chelation" as a basis for your decision, which of the following does not belong?
    1. penicillamine
    2. lead paint
    3. medroxyprogesterone
    4. heavy metal
155. Coadministration of terfenadine and which of the following juices may produce an unwanted result?
    1. grapefruit juice
    2. tomato juice
    3. cranberry juice
    4. apple juice
156. Which of the following is a common side effect reported with colchicine?
    1. headache
    2. flushes
    3. nausea and vomiting
    4. rash
157. Which is a trade name for verapamil?
    1. PROCARDIA
    2. NIMOTOP
    3. VERELAN
    4. none of the above
158. INDOCIN is to indomethacin, as ____ is to MINOCIN.
    1. minocycline
    2. minoxidil
    3. miconazole
    4. metronidazole
159. What do fluoxetine, sertraline, and paroxetine have in common?
    1. MAO inhibitors
    2. ACE inhibitors
    3. SSR inhibitors
    4. proton pump inhibitors
160. What information should you give the patient taking ketorolac?
    1. Watch for and report signs of gastrointestinal adverse events.
    2. Limit use of oral form to 5 days.
    3. Double dose for enhanced pain relief.
    4. Take aspirin to boost the effectiveness.
161. Coadministration of nifedipine and which of the following may result in a potentially serious or even life-threatening drug interaction?
    1. albuterol
    2. bumetanide

3. cimetidine
4. dextromethorphan

162. Which of the following is the trade name for ranitidine?
    1. ZOFRAN
    2. ZANTAC
    3. ZOCOR
    4. ZOLOFT

163. Using the key phrase, "thyroid gland" as the basis for your decision, which of the following does not belong?
    1. levothyroxin
    2. iodine
    3. idoxuridine
    4. goiter

164. Which of the following is the generic name for MEVACOR?
    1. lovastatin
    2. pravastatin
    3. simvastatin
    4. gemfibrozil

165. ACE inhibitors are contraindicated in patients treated with which of the following?
    1. prednisone
    2. potassium products
    3. potassium-sparing diuretics
    4. answers 2 and 3

166. Which of the following is a trade name for penicillin VK?
    1. VEETIDS
    2. VALIUM
    3. VASOTEC
    4. PENTAM

167. Which of the following drug classes or groups is represented by captopril?
    1. cardiovasculars (ACE inhibitors)
    2. cardiovasculars (calcium channel blockers)
    3. cardiovasculars (vasodilators)
    4. respiratory agents (beta antagonists)

168. How would you classify or group verapamil?
    1. ACE inhibitor
    2. Beta-adrenergic blocker
    3. inotropic agent
    4. calcium channel blocker

169. Which of the following answers best describes the primary indicated use for loratidine?
    1. angina
    2. hypercholesterolemia
    3. relieves symptoms of seasonal allergy
    4. edema

170. Which answer best describes the usual adult dose for ketorolac?
    1. 25 mg. QID
    2. maximum 40 mg./d
    3. 25 mg. TID
    4. 10 mg. qhs
171. What do itraconazole, fluconazole, and ketoconazole have in common?
    1. They are "azoles."
    2. They are antifungal agents.
    3. Coadministration with terfenadine is contraindicated.
    4. all of the above
172. Which of the following is a common side effect reported with ethinyl estradiol/levonorgestrel use?
    1. hair loss
    2. memory loss
    3. bloating
    4. tinnitus
173. Which of the following answers best describes a unique association with diclofenac?
    1. as an NSAID, has an unwanted impact on the gastric lining
    2. as an antibiotic, has an unwanted impact on destroying intestinal flora
    3. as an antineoplastic, has an unwanted impact on hair loss
    4. as an antihistamine, has an unwanted impact on drowsiness
174. Which of the following answers best describes the primary indicated use for triamterene/hydrochlorothiazide?
    1. mild to moderate pain
    2. edema in CHF
    3. rash
    4. elevated blood sugar
175. Which of the following is the generic name for SELDANE?
    1. astemizole
    2. loratidine
    3. clemastine
    4. terfenadine
176. Coadministration of lisinopril and which of the following may result in a potentially serious or even life-threatening drug interaction?
    1. lithium
    2. loperamide
    3. prochlorperazine
    4. glyburide
177. Digoxin is contraindicated in patients treated with which of the following?
    1. dexamethasone
    2. diphenoxylate
    3. phenazopyridine
    4. amiodarone

178. Using the key word, "ulceration" as the basis for your decision, which of the following does not belong?
    1. ibuprofen
    2. naproxen
    3. aspirin
    4. loperamide
179. What is the brand name for lovastatin?
    1. ZOCOR
    2. ZOLOFT
    3. MEVACOR
    4. PRAVACHOL
180. Which of the following is a side effect reported with verapamil?
    1. bulging eyes
    2. lightheadedness
    3. numbness
    4. tremor
181. Which of the following drug classes or groups is represented by phenytoin?
    1. antiseizure agents (dicarbamates)
    2. antiseizure agents (hydantoins)
    3. antiseizure agents (succinimides)
    4. psychotherapeutic agents (antidepressants)
182. Which of the following answers best describes the most common indication for amoxicillin?
    1. infection caused by susceptible strains of pathogens
    2. obsessive-compulsive disorders
    3. peptic ulcer, duodenal
    4. major depression
183. How would you classify or group ondansetron?
    1. antiemetic agent
    2. antineoplastic agent
    3. anti-inflammatory agent
    4. psychotherapeutic agent
184. Which answer best describes the usual adult therapeutic oral dose for enalapril used for hypertension?
    1. 10mg. to 40mg./day
    2. 20 mg. BID
    3. 20 mg. TID
    4. 2.5 mg. qhs
185. What information should you give the patient taking warfarin?
    1. Report black stools.
    2. Use a soft tooth brush.
    3. Protect the medication from light.
    4. all of the above
186. Which of the following is side effect reported with nifedipine?
    1. flushing
    2. hallucinations
    3. rash
    4. chills

Drugs

187. Which is an appropriate auxiliary label to use with nifedipine?
    1. TAKE WITH FOOD OR MILK (sustained release form)
    2. DO NOT CHEW, CRUSH, OR BREAK (sustained release form)
    3. TAKE MEDICATION WITH PLENTY OF WATER
    4. AVOID PROLONGED EXPOSURE TO SUNLIGHT
188. Potassium-wasting diuretics could be dangerous if used in patients treated with which of the following?
    1. TCAs
    2. NSAIDs
    3. proton pump inhibitors
    4. benzodiazepines
189. Coadministration of enalapril and which of the following may produce an unwanted result?
    1. diphenhydramine
    2. diphenylhydantoin
    3. liothyronine
    4. lithium
190. Using the key word "edema" as the basis for your decision, which of the following does not belong?
    1. nandrolone
    2. furosemide
    3. spironolactone
    4. chlorothiazide
191. Coadministration of triamterene/hydrochlorothiazide and which of the following may produce an unwanted result?
    1. ACE inhibitors
    2. lithium
    3. oral hypoglycemic drugs
    4. all of the above
192. Which of the following answers best describes the primary indicated use for penicillamine?
    1. gram positive bacterial infection
    2. chronic obstructive pulmonary disease
    3. heavy metal poisoning
    4. fungal dermatitis
193. Which is an appropriate auxiliary label to use with glipizide?
    1. TAKE WITH FOOD OR MILK
    2. DO NOT TAKE ASPIRIN
    3. DO NOT CRUSH
    4. MAY CAUSE DROWSINESS
194. Potassium-sparing diuretics are contraindicated in patients treated with which of the following?
    1. ACE inhibitors
    2. TCAs
    3. potassium preparations
    4. answers 1 and 3
195. Which of the following is the generic name for LEVOTHROID?
    1. levothyroxin
    2. levorphanol

3. levonorgestrel
4. levodopa

196. What action does cefprozil elicit?
    1. bacteriostatic
    2. bactericidal
    3. gene alteration
    4. pathogen replacement

197. Which answer best describes the usual adult dose for omeprazole?
    1. not to exceed 360 mg./day depending on condition
    2. maintain at 300 mcg./day for any condition
    3. 360 mcg. BID for all conditions
    4. 300 mcg. qhs for all conditions

198. What is the brand name for sertraline?
    1. ZOCOR
    2. ZANTAC
    3. SERZONE
    4. ZOLOFT

199. Which of the following is a common side effect reported with enalapril use?
    1. hallucinations
    2. bruising
    3. cough
    4. fever

200. NSAIDs are contraindicated in patients treated with which of the following?
    1. antiprotozoal agents
    2. antihistamine agents
    3. warfarin anticoagulants
    4. muscle relaxants

201. What information should you give the patient taking ethinyl estradiol/norethindrone?
    1. Discontinue if you are pregnant or intend to become pregnant.
    2. Double missed doses.
    3. Avoid acidic juices while taking this medication.
    4. Decrease dosage if experiencing speech disturbances.

202. Clarithromycin should never be used with which of the following?
    1. non-sedating antihistamines
    2. calcium channel blockers
    3. protease inhibitors
    4. coenzyme A reductase inhibitors

203. Using the key phrase, "serotonin reuptake inhibitor" as the basis for your decision, which of the following does not belong?
    1. sertraline
    2. fluoxetine
    3. desipramine
    4. paroxetine

204. How would you classify or group warfarin?
    1. antiarrhythmic agent
    2. anticoagulant agent

3. antianginal agent
4. antianxiety agent
205. Many oral antidiabetic drugs may be contraindicated in patients treated with which of the following?
    1. meperidine
    2. allopurinol
    3. phenylbutazone
    4. acetaminophen
206. What is/are the brand name(s) for methotrexate?
    1. FOLEX
    2. METICORTEN
    3. RHEUMATREX
    4. answers 1 and 3
207. Which of the following is the generic name for PRILOSEC?
    1. omeprazole
    2. olsalazine
    3. olfloxacin
    4. prazosin
208. Which of the following is a side effect reported with ondansetron?
    1. headache
    2. urinary retention
    3. bronchospasm
    4. excessive salivation
209. Which of the following drug classes or groups is represented by loratidine?
    1. antihistamines
    2. steroids
    3. hormones
    4. analgesics
210. Which of the following answers best describes the primary indicated use for warfarin?
    1. tachycardia
    2. heart palpitations
    3. blood clots
    4. bradycardia
211. Theophylline-type asthma drugs are contraindicated in patients treated with which of the following?
    1. vitamin B-12 supplements
    2. hyperosmotic laxatives
    3. antidiabetic agents
    4. quinolone antibiotics
212. Which answer best describes the usual adult oral dose for famotidine?
    1. 20 mg. TID
    2. 15 mg. QID
    3. maintenance dose up to 40 mg. per day
    4. depends on body weight
213. Which is a brand name for verapamil?
    1. PLENDIL
    2. ISOPTIN

3. DYNACIRC
4. CARDENE

214. BRETHINE is to terbutaline, as BRETHAIRE is to ____.
    1. isoetharine
    2. terbutaline
    3. epinephrine
    4. ephedrine

215. Which of the following is a common side effect reported with famotidine use?
    1. dizziness
    2. headache
    3. change in bowel movements
    4. all of the above

216. What information should you give the patient taking nabumetone?
    1. Report blackened stools.
    2. Reduce dose to relieve stomach disturbances.
    3. Increase dose for increased effect.
    4. Take with orange juice.

217. Non-sedating antihistamines are contraindicated in patients treated with which of the following?
    1. macrolide antibiotics
    2. antiviral agents
    3. "Azole" antifungal drugs
    4. answers 1 and 3

218. Which is a side effect reported with warfarin use?
    1. blurred vision
    2. hair loss
    3. thirst
    4. tinnitus

219. Which is an appropriate strip label to use with paroxetine?
    1. MAY CAUSE DROWSINESS
    2. DO NOT CRUSH OR CHEW MEDICATION
    3. AVOID ALCOHOL
    4. answers 1 and 3

220. Cisapride can have a potentially fatal interaction with which of the following?
    1. multiple vitamins containing minerals
    2. antifungal products
    3. analgesic products
    4. psychotherapeutic products

221. Using the key phrase, "fungal infection" as the basis for your decision, which of the following does not belong?
    1. fluconazole
    2. metronidazole
    3. clotrimazole
    4. ketaconazole

222. What is the trade name for famotidine?
    1. PEPCID
    2. PROPULSID

3. PRILOSEC
4. AXID

223. Serotonin-type antidepressants are contraindicated in patients treated with which of the following?
    1. NSAIDs
    2. MAO inhibitors
    3. proton pump inhibitors
    4. HMG-CoA reductase inhibitors

224. Which of the following is the generic name for GLUCOTROL?
    1. glyburide
    2. tolbutamide
    3. glucagon
    4. glipizide

225. Coadministration of verapamil and which of the following may produce an unwanted result?
    1. quinidine
    2. carbamazapine
    3. cimetidine
    4. all of the above

226. Which of the following best describes the drug class for azathioprine?
    1. cardiovascular agents
    2. anticonvulsant drugs
    3. nonsteroidal anti-inflammatory agents
    4. immunosuppressive agents

227. What notable information should you know about ondansetron?
    1. All forms of the medication must be kept in the refrigerator.
    2. Maximum daily dose of 8 mg. is used to PREVENT and not to TREAT nausea and vomiting.
    3. Give dose 30 minutes before chemotherapy dosing.
    4. answers 2 and 3

228. Which of the following answers best describes the most common indication for nifedipine?
    1. depression
    2. gastric hypersecretion
    3. angina
    4. myocardial infarction

229. Hydantoin anticonvulsant agents are contraindicated in patients treated with which of the following?
    1. cyclosporine
    2. bumetanide
    3. glipizide
    4. sucralfate

230. Which of the following is a common side effect reported with digoxin use?
    1. loss of appetite
    2. hair loss
    3. excessive hair growth
    4. coating on the tongue

231. Which of the following is the trade name for simvastatin?
    1. PRAVACHOL
    2. MEVACOR
    3. lOPID
    4. ZOCOR

232. What information should you give the patient taking naproxen?
    1. Avoid foods that can cause heartburn.
    2. Double up on missed doses.
    3. Decrease dose if stools are blackened.
    4. Rise slowly from a seated or lying position.

233. While taking captopril, special care should be taken to monitor concurrent use of which of the following drugs?
    1. potassium-sparing diuretics
    2. NSAIDs
    3. oral contraceptives
    4. muscle relaxants

234. The "azole" in the names of fluconazole, ketaconazole, miconazole, and clotrimazole best describe these drugs as which of the following?
    1. antiviral
    2. antibacterial
    3. antifungal
    4. antiprotozoal

235. Carbamazepine anticonvulsant agent is contraindicated in patients treated with which of the following?
    1. macrolide antibiotics
    2. propoxyphene analgesics
    3. saline laxatives
    4. answers 1 and 2

236. Coadministration of paroxetine and which of the following may result in a potentially serious or even life-threatening drug interaction?
    1. salt replacement products
    2. MAO inhibitors
    3. ethanolamine antihistamines
    4. fluoroquinolones

237. Which of the following is a trade name for insulin?
    1. DIABETA
    2. MICRONASE
    3. GLUCOTROL
    4. HUMULIN

238. Which is an appropriate strip label to use with diclofenac?
    1. TAKE ON AN EMPTY STOMACH
    2. TAKE WITH FOOD OR MILK
    3. DO NOT TAKE DAIRY PRODUCTS OR ANTACID PREPARATIONS WITHIN ONE HOUR OF THIS MEDICATION
    4. FINISH ALL THIS MEDICATION UNLESS OTHERWISE DIRECTED BY PRESCRIBER

Drugs

239. What is the generic name for the drug DILANTIN?
    1. primidone
    2. phenytoin
    3. valproic acid
    4. mephenytoin

240. How would you classify or group ibuprofen?
    1. antiarrhythmic agent
    2. nonsteroidal anti-inflammatory drug
    3. antihistamine
    4. central nervous system stimulant

241. Anticoagulant agents are contraindicated in patients treated with which of the following?
    1. guaifenesin
    2. kaolin/Pectin
    3. metronidazole
    4. loperamide

242. Which of the following answers best describes the primary indicated use for digoxin?
    1. PUD
    2. GERD
    3. CHF
    4. NIDDM

243. Which answer best describes the usual adult oral dose for fluoxetine?
    1. maximum 80 mg. per day divided in the morning and at noon
    2. 20 mg. TID
    3. 20 mg. QID
    4. 10 mg. as needed

244. Using the key word, "neurotransmission" as the basis for your decision, which of the following does not belong?
    1. sertraline
    2. paroxetine
    3. fluoxetine
    4. cephradine

245. What is the trade name for penicillamine?
    1. COUMADIN
    2. DEPEN
    3. ELDEPRYL
    4. FLEXERIL

246. Which is an appropriate auxiliary label to use with levothyroxin?
    1. AVOID SMOKING WHILE TAKING THIS MEDICATION
    2. USE THIS MEDICATION EXACTLY AS DIRECTED. DO NOT SKIP DOSES OR DISCONTINUE UNLESS DIRECTED BY YOUR DOCTOR
    3. DO NOT CRUSH
    4. TAKE WITH FOOD OR MILK

247. Co-trimoxazole compounds are contraindicated in patients treated with which of the following?
    1. multiple vitamins containing trace elements
    2. anticoagulants

3. glucocorticosteroids
4. NSAIDs

248. Coadministration of clonazepam and which of the following may result in a potentially serious or even life-threatening drug interaction?
    1. prednisone
    2. nefazodone
    3. ibuprofen
    4. captopril

249. Using the key acronym, "PUD" as the basis for your decision, which of the following does not belong?
    1. nizatidine
    2. famotidine
    3. cisapride
    4. sucralfate

250. Which of the following is the trade name for nizatidine?
    1. AZLIN
    2. ATIVAN
    3. AXID
    4. NIZORAL

251. Which of the following is the generic name for CYTOXAN?
    1. cyclophosphamide
    2. cyclobenzaprine
    3. cycloserine
    4. cyclosporine

252. Which of the following is the generic name for ZOCOR?
    1. simvastatin
    2. pravastatin
    3. probucol
    4. gemfibrozil

253. Benzodiazepines are contraindicated in patients treated with which of the following?
    1. macrolide antibiotics
    2. HMG-CoA reductase inhibitors
    3. bulk laxatives
    4. hormones

254. Which of the following drug classes or groups is represented by fluoxetine?
    1. NSAID
    2. beta Blocker
    3. psychotherapeutic agent
    4. steroid

255. Which of the following answers best describes the most common indications for conjugated estrogens?
    1. contraception
    2. sexually transmitted disease treatment
    3. estrogen replacement therapy for menopausal symptoms and osteoporosis
    4. conjunctivitis

256. Which answer best describes the usual adult oral dose for dicofenac?
    1. 200 mg. per day in 2–5 divided doses
    2. 200 mg. TID
    3. up to 1000 mg. per day
    4. 1000 mg. per day in 4 equally divided doses
257. Which of the following is a common side effect reported with fluconazole use?
    1. rash
    2. flushing
    3. hallucinations
    4. numbness in toes and fingers
258. Coadministration of ibuprofen and which of the following may result in a potentially serious or even life-threatening drug interaction?
    1. fluoxetine
    2. lithium
    3. diphenhydramine
    4. diphenylhydantoin
259. How would you classify or group paclitaxel?
    1. antihistamine agent
    2. antiulcer agent
    3. antineoplastic agent
    4. anti-infective agent
260. The "idine" in the names of cimetidine, famotidine, nizatidine, ranitidine best describe these drugs as which of the following?
    1. beta blockers
    2. histamine H2 antagonists
    3. NSAIDs
    4. steroids
261. Which of the following is the trade name for levothyroxin?
    1. SYMMETREL
    2. SYNALGOS
    3. LEVSIN
    4. SYNTHROID
262. Which of the following is a trade name for potassium chloride?
    1. KAOPECTATE
    2. KLONOPIN
    3. KAY CIEL
    4. KEFLEX
263. Which of the following drug classes or groups is represented by cefaclor?
    1. cardiovascular (beta blocker)
    2. analgesic (opiod)
    3. antibiotic (cephalosporin)
    4. bronchodilator (xanthine)
264. Which of the following answers best describes the primary indicated use for cyclophosphamide?
    1. glaucoma
    2. muscle spasm

3. carcinomas
4. allergic symptoms

265. Which of the following answers best describes the primary indicated use for ondansetron?
    1. diarrhea resulting from protozoal infestation
    2. migraine headache
    3. emesis associated with emetogenic chemotherapeutic agents
    4. chronic fatigue syndrome

266. Which answer best describes the usual adult dose for lisinopril?
    1. up to 40 mcg. per day
    2. up to 40 mg. per day
    3. 40 mcg. BID
    4. 40 mg. BID

267. Which of the following is side effect reported with nizatidine?
    1. cough
    2. runny nose
    3. sweats
    4. constipation

268. What principle information should be noted with alprozolam?
    1. Avoid alcohol.
    2. may cause drug dependence
    3. Discontinue gradually after prolonged use.
    4. all of the above

269. Coadministration of conjugated estrogens and which of the following may produce an unwanted result?
    1. pantothenic acid
    2. folic acid
    3. ascorbic acid
    4. nalidixic acid

270. Which of the following is the trade name for ethinyl estradiol/norethindrone?
    1. TRIPHASIL
    2. OGEN
    3. NORDETTE
    4. ORTHO-NOVUM

271. Which answer best describes the usual adult dose for prazosin?
    1. maximum dose of 20 mcg./d
    2. minimum dose of 20 mg./d
    3. 20 mcg./Kg. TID
    4. maximum dose of 20 mg./d

272. COLESTID is to colestipol as COLY-MYCIN S is to ____.
    1. colfosceril
    2. colchicine
    3. colistin
    4. clofibrate

273. How would you classify or group omeprazole?
    1. selective serotonin reuptake inhibitor
    2. proton pump inhibitor

3. selective 5–HT 3 receptor antagonist
4. TCA

274. Which of the following answers best describes the primary indicated use for ibuprofen?
    1. mild to moderate pain and inflammation associated with inflammatory diseases
    2. tachycardia
    3. symptoms associated with seasonal allergy
    4. depression

275. Which answer best describes the usual adult oral dose for glyburide?
    1. up to 40 mg. per day
    2. up to 30 mg. per day
    3. 10 mg. TID
    4. up to 20 mg. per day

276. Which of the following is a side effect reported with levothyroxin?
    1. tachycardia
    2. nervousness
    3. changes in menstrual cycle
    4. all of the above

277. What information should you give the patient taking conjugated estrogens?
    1. Report unusual bleeding or staining.
    2. Report experiences of depression.
    3. Do not take this medication if pregnant or intend to become pregnant.
    4. all of the above

278. Which of the following is a significant side effect reported with paclitaxel?
    1. hypotension
    2. loss of hair
    3. thirst
    4. wheezing

279. Using the key phrase, "birth control" as the basis for your decision, which of the following does not belong?
    1. mestranol/norethindrone
    2. mestranol/norethynodrel
    3. ethinyl estradiol/levonorgestrel
    4. estradiol

280. Which of the following is the trade name for glyburide?
    1. DIABETA
    2. GLYNASE
    3. MICRONASE
    4. all of the above

281. Which of the following is the generic name for TORADOL?
    1. ketoprofen
    2. ketorolac
    3. ketamine
    4. tolmetin

282. Clindamycin is to CLEOCIN as ____ is to CLINORIL?
    1. sulindac
    2. clemastine
    3. sumatriptan
    4. clarithromycin
283. Which of the following is the trade name for verapamil?
    1. CALAN
    2. CARDIZEM
    3. NORVASC
    4. VASCOR
284. How would you classify or group nizatidine?
    1. SSRI
    2. histamine H-2 blocker
    3. selective 5-HT-3 receptor blocker
    4. NSAID
285. Which is an appropriate auxiliary label to use with phenytoin?
    1. TAKE ON AN EMPTY STOMACH
    2. MAY CAUSE DROWSINESS
    3. DO NOT CRUSH OR CHEW
    4. TAKE WITH PLENTY OF WATER
286. Which of the following answers best describes the primary indicated use for diclofenac?
    1. bacterial infection
    2. relieve pain and reduce inflammation
    3. carcinoma
    4. reduce cholesterol level
287. Which answer best describes the usual adult oral dose for digoxin?
    1. up to 0.5 mg. per day
    2. 1.5 mg. loading dose, 3.0 mg. daily, thereafter
    3. 0.125 mg. QID
    4. 0.25 mg. Q.O.D.
288. Which of the following is a common side effect reported with diltiazem?
    1. vomiting
    2. nausea
    3. headache
    4. constipation
289. Which of the following answers best describes a unique association with glyburide?
    1. increased risk of ototoxicity
    2. increased risk of prostate exacerbation
    3. increased risk of cardiovascular mortality
    4. increased risk of impotence
290. Coadministration of fluoxetine and which of the following may produce an unwanted result?
    1. diet high in vitamin K
    2. glipizide
    3. ibuprofen
    4. MAO inhibitors

291. What information should you give the patient taking paroxetine?
    1. A response may take several weeks.
    2. Doses should be adjusted at 7-day intervals.
    3. Take medication with food.
    4. answers 1 and 2
292. Using the key word, "hypertension" as the basis for your decision, which of the following does not belong?
    1. isradipine
    2. diltiazem
    3. enalapril
    4. procainamide
293. Which of the following is the generic name for ZESTRIL?
    1. verapamil
    2. ramipril
    3. lisinopril
    4. bepridil
294. How would you classify or group methotrexate?
    1. antirheumatic agent
    2. antipsoriatic agent
    3. antineoplastic agent
    4. all of the above
295. Which of the following answers best describes the primary indicated use for gold sodium thiomalate?
    1. delayed puberty
    2. neurogenic diabetes insipidus
    3. inhibited growth
    4. rheumatoid arthritis
296. Which answer best describes the usual adult dose for levothyroxin?
    1. up to 200 mg. per day
    2. 100 mg. BID
    3. up to 200 mcg. per day
    4. 200 mcg. BID
297. Coadministration of phenytoin and which of the following may result in a potentially serious or even life-threatening drug interaction?
    1. loperamide
    2. ipratropium
    3. cimetidine
    4. penicillin
298. Which of the following side effects is commonly reported with fluoxetine use?
    1. hair dryness
    2. dry mouth
    3. hoarseness
    4. sore throat
299. Which is an appropriate strip label to use with naproxen?
    1. TAKE WITH FOOD OR MILK
    2. TAKE ON AN EMPTY STOMACH

3. AVOID EXPOSURE TO SUNLIGHT
4. none of the above

300. Coadministration of methotrexate and which of the following may result in a potentially serious or even life-threatening drug interaction?
    1. analgesics
    2. ACE inhibitors
    3. NSAIDs
    4. antispasmodics

301. The "pril" in the names of captopril, enalapril, lisinopril, and ramipril best describes these drugs as which of the following?
    1. calcium channel blockers
    2. ACE inhibitors
    3. inotropics
    4. beta-adrenergic blockers

302. What is the trade name for nabumetone?
    1. REGLAN
    2. RELAFEN
    3. RESTORIL
    4. NALFON

303. Using the key phrase, "cardiac arrhythmias" as the basis for your decision, which of the following does not belong?
    1. quinidine
    2. diphenhydramine
    3. procainamide
    4. diphenylhydantoin

304. Which of the following drugs is a protease inhibitor?
    1. acyclovir
    2. stavudine
    3. zidovudine
    4. indinavir

305. Which of the following answers best describes the primary indicated use for hydroxychloroquine?
    1. malaria
    2. muscle spasm
    3. depression
    4. myocardial infarction

306. Which answer best describes the usual adult dose for naproxen?
    1. 1.5 Gm./d
    2. 1.5 mg./d
    3. 1.5 mcg. QID
    4. 1.5 Gm. BID

307. Which of the following is a common side effect reported with furosemide?
    1. vomiting
    2. orthostatic hypotension
    3. nausea
    4. dehydration

Drugs

308. What information should you give the patient taking nizatidine?
    1. Avoid aspirin.
    2. Avoid black pepper, caffeine, and harsh spices.
    3. Avoid alcohol.
    4. all of the above
309. Which is a brand name for penicillamine?
    1. CUPRIMINE
    2. COUMADIN
    3. CRIXIVAN
    4. PEN-VEE K
310. Coadministration of medroxyprogesterone and which of the following may produce an unwanted result?
    1. aminoglutethimide
    2. bumetamide
    3. chlorpropamide
    4. desonide
311. Which answer best describes the usual adult dose for pravastatin?
    1. 40 mg. TID
    2. up to 40 mg./day
    3. 40 mcg./Kg./d in divided doses
    4. up to 40 mcg. once a day
312. Using the key words, "calcium channel blocker" as the basis for your decision, which of the following does not belong?
    1. metoprolol
    2. felodipine
    3. bepridil
    4. diltiazem
313. Which of the following is the trade name for naproxen?
    1. NAPROSYN
    2. ANAPROX
    3. NALFON
    4. answers 1 and 2
314. What information should you give the patient taking enalapril?
    1. Report a persistent cough.
    2. Do not crush tablets.
    3. Take medication with food or milk.
    4. Take medication on an empty stomach.
315. Which of the following answers best describes the primary indicated use for furosemide?
    1. dermatitis
    2. edema
    3. arrhythmia
    4. none of the above
316. Which of the following is the generic for ZOLOFT?
    1. sertraline
    2. selegiline
    3. simvastatin
    4. simethicone

317. Which answer best describes the usual adult dose for nizatidine?
    1. 300 mcg. qhs
    2. 300 mg. BID
    3. up to 300 mg./day
    4. 30 mg. prn
318. Which of the following is a side effect reported with gold sodium thiomalate?
    1. metallic taste
    2. tongue irritation
    3. oral ulcers
    4. all of the above
319. Which is an appropriate auxiliary label to use with nizatidine?
    1. AVOID PROLONGED EXPOSURE TO SUNLIGHT
    2. MAY CAUSE DROWSINESS
    3. TAKE ON AN EMPTY STOMACH
    4. PROTECT MEDICATION FROM LIGHT
320. Coadministration of gold sodium thiomalate and which of the following may produce an unwanted result?
    1. pentazocine
    2. pentobarbital
    3. pentoxifylline
    4. penicillamine
321. The "dipine" in the names of felodipine, isradipine, nicardipine, nifedipine, and nimodipine best describes these drugs as which of the following?
    1. ACE inhibitors
    2. calcium channel blockers
    3. beta-adrenergic blockers
    4. alpha-adrenergic blockers
322. How would you classify or group paroxetine?
    1. antidiarrheal
    2. antidepressant
    3. antiulcer
    4. analgesic
323. Which is a brand name for nifedipine?
    1. PROCAN
    2. PROCARDIA
    3. PRINVIL
    4. NAVANE
324. What is the generic name for NAPROSYN?
    1. nandrolone
    2. nicardipine
    3. naloxone
    4. naproxen
325. What is the trade or brand name for the generic drug ciprofloxacin?
    1. NOROXIN
    2. FLOXIN
    3. MAXAQUIN
    4. CIPRO

326. Which of the following is the trade name for warfarin?
    1. CALAN
    2. WYDASE
    3. COUMADIN
    4. CAPOTEN
327. Amlodipine is to NORVASC, as ____ is to NAVANE.
    1. amoxapine
    2. thiothixene
    3. amitriptyline
    4. thioridazine
328. Which of the following is a side effect reported with ibuprofen?
    1. diarrhea
    2. heartburn
    3. constipation
    4. irritability
329. Which of the following answers best describes the primary indicated use for paclitaxel?
    1. malignant prostate cancer
    2. lymphoma
    3. testicular carcinoma
    4. carcinoma of the ovaries and breast
330. What information should you give the patient taking gold sodium thimalate?
    1. Do not stand near microwave ovens.
    2. Loss of hair is common.
    3. Benefits of gold therapy may require at least 3 months.
    4. Expect excessive frequency of urination.
331. Coadministration of lovastatin and which of the following may result in a potentially life threatening drug interaction?
    1. macrolide antibiotics
    2. cholesterol drugs
    3. transplant drugs
    4. all of the above
332. Using the key word, "inotropic" as the basis for your decision, which of the following does not belong?
    1. digitoxin
    2. metronidazole
    3. dosoxin
    4. milrinone
333. Which of the following is the trade name for fluoxetine?
    1. PROVERA
    2. PROVENTIL
    3. FIORINAL
    4. PROZAC
334. What is the trade or brand name for the generic drug digoxin?
    1. LOTENSIN
    2. LOPRESSOR
    3. LANOXIN
    4. LASIX

335. Which answer best describes the usual adult dose for propoxyphene napsylate/acetaminophen?
    1. starting dose of 100 mg. titrated up until pain is relieved
    2. maximum up to 600 mg./day
    3. 600 mcg. qhs
    4. 60 mg./ Kg. q6h
336. How would you classify or group insulin?
    1. immunosuppressive agent
    2. hypoglycemic agent
    3. osteoporosis agent
    4. colony stimulating factor
337. Which of the following answers best describes the most common indication for cefaclor?
    1. major depression
    2. seasonal affective disorder
    3. infection caused by susceptible organisms
    4. ulcers
338. Which answer best describes the usual adult oral dose for glipizide?
    1. 20 mg. per day
    2. up to 40 mg. per day
    3. 20 mg. TID
    4. 10 mg. QID
339. What information should you give the patient taking diltiazem?
    1. Discontinue gradually.
    2. Double the dose if you feel your heart pounding.
    3. Crush the sustained release form if you cannot swallow it whole.
    4. none of the above
340. Which of the following is the generic name for MICRONASE?
    1. glipizide
    2. glyburide
    3. tolazamide
    4. tolbutamide
341. Which of the following is a side effect reported with paroxetine?
    1. ejaculatory disturbance
    2. increased pulse rate
    3. fever
    4. yellowing skin
342. Using the key word, "antineoplastic" as the basis for your decision, which of the following does not belong?
    1. cyclophosphamide
    2. etoposide
    3. allopurinol
    4. melphalan
343. Which of the following is the trade name for diclofenac?
    1. VOSOL
    2. DIDRONEL
    3. VOLTAREN
    4. VIVACTIL

Drugs

344. What is the generic name for the drug LANOXIN?
    1. digoxin
    2. digitoxin
    3. diphenhydramine
    4. propranolol
345. How would you classify or group naproxen?
    1. TCA
    2. hormone
    3. NSAID
    4. ACE inhibitor
346. Which is an appropriate strip label to use with diltiazem?
    1. TAKE WITH MILK OR FOOD
    2. THIS MEDICATION MAY IMPAIR THE ABILITY TO DRIVE OR OPERATE MACHINERY
    3. TAKE WITH A FULL GLASS OF WATER
    4. MAY CAUSE DISCOLORATION OF THE URINE OR FECES
347. Which is an appropriate auxiliary label to use with potassium chloride?
    1. SWALLOW TABLETS WHOLE
    2. DILUTE BEFORE USING (liquid form)
    3. PROTECT FROM LIGHT
    4. answers 1 and 2
348. Which answer best describes the usual adult dose for lovastatin?
    1. maximum 8 Gm./day
    2. 80 mcg. qd
    3. 40 mg. qd
    4. maximum 80 mg./day
349. Which of the following is a side effect reported with insulin?
    1. hair loss
    2. loss of memory
    3. sweats
    4. metallic taste
350. Coadministration of sertraline and which of the following may result in a potentially serious or even life-threatening drug interaction?
    1. proton pump inhibitors
    2. ACE inhibitors
    3. MAO inhibitors
    4. histamine H-2 blockers
351. What is the trade or brand name for the generic drug amoxicillin?
    1. OMNIPEN
    2. UNASYN
    3. ERYTHROCIN
    4. AMOXIL
352. How would you classify or group lovastatin?
    1. antiemetic agent
    2. antidiarrheal agent
    3. anticonvulsant
    4. antihyperlipidemic agent

353. What information should you give the patient taking phenytoin?
    1. Use the same manufacturer's brand of drug.
    2. Take medication with food.
    3. Medication may cause drowsiness.
    4. all of the above

354. What information should you give the patient taking fluconazole?
    1. Complete the full course of therapy.
    2. Do not chew or crush the medication.
    3. Visual disturbances are common, but not dangerous.
    4. Increase your dietary fiber.

355. Which of the following answers best describes the usual dosage for cefaclor?
    1. children: no dose under 12 years old. Adult: 500 mg. BID
    2. children: 250 mg. TID. Adults: 500 mg. TID
    3. children and adult dose are the same
    4. children: up to 40 mg./Kg./day divided every 8–12 hours. Adults: 250–500 mg. q8h

356. Which of the following is a side effect reported with ketorolac?
    1. Loss of appetite
    2. joint pain
    3. loss of muscle coordination
    4. drowsiness

357. What information should you give the patient taking omeprazole?
    1. Take medication before eating.
    2. Swallow medication whole.
    3. Do not chew, crush, or break medication.
    4. all of the above

358. Long term concurrent use of acetaminophen and which of the following poses a potential threat for liver damage?
    1. alcohol
    2. anticonvulsant drugs
    3. tuberculosis drugs
    4. all of the above

359. Coadministration of potassium chloride and which of the following may result in a potentially serious or even life-threatening drug interaction?
    1. hypoglycemic drugs
    2. potassium-sparing diuretics
    3. laxatives
    4. antigout drugs

360. Using the key word, "inflammation" as the basis for your decision, which of the following does not belong?
    1. diflunisal
    2. fluconazole
    3. fenoprofen
    4. diclofenac

361. How would you classify or group lisinopril?
    1. inotropic agent
    2. calcium channel blocker
    3. ACE inhibitor
    4. beta-adrenergic blocker
362. How would you classify or group gold sodium thiomalate?
    1. nonsteroidal anti-inflammatory agent
    2. osteoporosis agent
    3. rheumatoid arthritic agent
    4. skeletal growth factor
363. What information should you give the patient taking sertraline?
    1. Dose changes may be made at 1-week intervals.
    2. Males may experience sexual dysfunction.
    3. An observable response may take several weeks.
    4. all of the above
364. Which of the following is the trade name for diltiazem?
    1. CARDIZEM
    2. TENORMIN
    3. ISOPTIN
    4. PLENDIL
365. Using the key word, "electrolyte" as the basis for your decision, which of the following does not belong?
    1. sodium chloride
    2. copper sulfate
    3. calcium gluconate
    4. potassium chloride
366. Which of the following is a side effect reported with nabumetone?
    1. blurred vision
    2. heartburn
    3. weakness
    4. thirst
367. What information should you give the patient taking glipizide?
    1. Do not skip meals.
    2. Do not take double doses.
    3. Avoid OTC cough/cold, appetite control, hayfever, and asthma preparations.
    4. all of the above
368. Coadministration of insulin and which of the following may produce an unwanted result?
    1. corticosteroids
    2. thyroid hormones
    3. oral contraceptives
    4. all of the above
369. Using the key word, "hormone" as the basis for your decision, which of the following does not belong?
    1. methyltestosterone
    2. methylprednisolone

3. insulin
4. estrogen

370. A representative of the histamine H-2 receptor antagonist class of drugs is which of the following?
    1. ranitidine
    2. ipratropium
    3. labetolol
    4. glipizide

371. Using the key phrase, "potassium-sparing" as the basis for your decision, which of the following does not belong?
    1. spironolactone
    2. triamterene/hydrochlorothiazide
    3. furosemide
    4. amiloride

372. What information should you give the patient taking digoxin?
    1. Report blood in stools.
    2. Report skin rash.
    3. Report discoloration of the tongue.
    4. Report visual disturbances.

373. Which of the following answers best describes the usual adult oral dosage for clarithromycin?
    1. Duration and dosage depend on the type of infection.
    2. 250 mg. TID for 21 days
    3. 500 mg. TID for 10–14 days
    4. dosage range between 250–500 mg. TID for 14 days

374. Which of the following is a side effect reported with omeprazole?
    1. glossitis
    2. arthralgia
    3. rapid heart rate
    4. changes in bowel movements

375. What is the generic name for the drug ZANTAC?
    1. ranitidine
    2. ramipril
    3. rifampin
    4. ritodrine

376. What is the trade name for ondansetron?
    1. ZYLOPRIM
    2. ZOFRAN
    3. ONCOVIN
    4. ZOSYN

377. What information should you give the patient taking lovastatin?
    1. Use medication in conjunction with low fat dietary restrictions.
    2. Take the medication with a full glass of water.
    3. Maintain a low salt diet.
    4. Take the medication during the evening.

378. Coadministration of ethinyl estradiol/norethindrone and which of the following may produce an unwanted result?
    1. metoprolol
    2. phenytoin

Drugs

3. theophylline
4. all of the above

379. Using the key word, "gout" as the basis for your decision, which of the following does not belong?
    1. colchicine
    2. probucol
    3. allopurinol
    4. probenecid

380. Which of the following is the trade name for enalapril?
    1. VASODILAN
    2. VASOCIDIN
    3. VASCOR
    4. VASOTEC

381. Which is an appropriate auxiliary label to use with hydroxychloroquine?
    1. TAKE WITH FOOD OR MILK
    2. AVOID SMOKING WHILE TAKING THIS MEDICATION
    3. DO NOT CRUSH
    4. TAKE ON AN EMPTY STOMACH

382. Which of the following answers best describes the primary indicated use for glipizide?
    1. noninsulin-dependent diabetes mellitus
    2. insulin-dependent diabetes mellitus
    3. juvenile diabetes
    4. none of the above

383. Which is an appropriate strip label to use with triamterene/hydrochlorothiazide?
    1. TAKE ON AN EMPTY STOMACH
    2. AVOID PROLONGED EXPOSURE TO SUNLIGHT
    3. TAKE AFTER MEALS
    4. answers 2 and 3

384. Which of the following answers best describes the usual oral dosage for conjugated estrogens?
    1. 0.3 mg. twice a day
    2. up to 7.5 mg. daily depending on indication
    3. the highest dose tolerated for 20 days
    4. 1.25 mg. daily for 7 days, followed by 0.625 mg. for 21 days

385. Which is an appropriate auxiliary label to use with omeprazole?
    1. MAY CAUSE DROWSINESS
    2. DO NOT CRUSH. SWALLOW WHOLE
    3. MAY CAUSE DISCOLORATION OF URINE OR FECES
    4. AVOID PROLONGED EXPOSURE TO SUNLIGHT

386. Allopurinol given with which of the following drug groups may precipitate a potentially fatal reaction called the Stevens-Johnson Syndrome?
    1. ACE inhibitors
    2. corticosteroids
    3. serotonin reuptake inhibitors
    4. diuretics

387. Which of the following answers best describes a unique association with fluoxetine?
    1. This medication is used as an adjunct to TCAs.
    2. An observable response may take several weeks.
    3. The drug has an extremely high safety profile.
    4. The effectiveness of this medication can be measured only through blood tests.
388. Using the key word, "seizures" as the basis for your decision, which of the following does not belong?
    1. diphenhydramine
    2. clonazepam
    3. phenytoin
    4. valparoic acid
389. What is the brand name for gold sodium thiomalate?
    1. MYLERAN
    2. MYOCHRISINE
    3. MYSOLINE
    4. MYCOBUTIN
390. What is the generic name for AXID?
    1. nifedipine
    2. nizatidine
    3. niacin
    4. azathioprine
391. Amitriptyline is to aminophylline as ____ is to bronchodilator.
    1. antibiotic
    2. antineoplastic
    3. antihypertensive
    4. antidepressant
392. Which of the following is a primary indication/use for colchicine, while hypercholesterolemia is the primary indication/use for the similarly sounding drug, colestipol?
    1. gouty arthritis
    2. viral infection
    3. diarrhea
    4. migraine
393. Which of the following answers best describes the primary indicated use for paroxetine?
    1. peptic ulcer disease (PUD)
    2. myocardial infarction (MI)
    3. depression
    4. asthma
394. Which of the following is a side effect reported with lisinopril?
    1. cough
    2. itching
    3. nausea and vomiting
    4. excessive thirst
395. What information should you give the patient taking methotrexate?
    1. Cranberry juice will relieve the burning urination.
    2. Avoid alcohol.

Drugs

3. Stay out of sunlight while on this medication.
4. answers 2 and 3

396. What information should you give the patient taking levothyroxin?
   1. Take each dose with a full glass of orange juice.
   2. The drug may cause drowsiness.
   3. Noticeable effects may take a few weeks.
   4. Double up on missed doses.

397. Which of the following drugs can potentially interact with azathioprine?
   1. propranolol
   2. haloperidol
   3. allopurinol
   4. terconazole

398. QUINAMM is to QUINORA as ____ is to quinidine.
   1. quinapril
   2. quinacrine
   3. quinine
   4. quazepam

399. Which answer best describes the usual adult dose for ranitidine?
   1. 300 mg. daily
   2. 300 mg. BID
   3. 150 mg. TID
   4. 150 mg. QID

400. What test antigens are included in an anergy panel or anergy testing kit?
   1. candida albicans skin test antigen, mumps skin test antigen, tuberculin test
   2. candida albicans skin test antigen, poison ivy skin test antigen, histoplasmin
   3. coccidioidin, histoplasmin, contact dermatitis skin test antigen, tuberculin test
   4. candida albicans skin test antigen, mumps skin test antigen, electrical skin test antigen

401. What is the trade or brand name for generic conjugated estrogens?
   1. OGEN
   2. PROVERA
   3. OVRAL
   4. PREMARIN

402. Which is an appropriate auxiliary label to use with insulin?
   1. REFRIGERATE AND SHAKE WELL BEFORE USING
   2. FOR EXTERNAL USE
   3. TAKE ON AN EMPTY STOMACH 1 HOUR BEFORE OR 2 TO 3 HOURS AFTER A MEAL
   4. DILUTE BEFORE ADMINISTRATION

403. What information should you give the patient taking insulin?
   1. Do not use if there are clumps in the insulin after mixing or if bottle has a frosted appearance.
   2. Exercise may lower your need for insulin.

3. Illness may cause your insulin requirements to change.
4. all of the above

404. Which is an appropriate strip label to use with digoxin?
    1. MAY CAUSE DISCOLORATION OF URINE OR FECES
    2. TAKE WITH PLENTY OF WATER
    3. DRINK A FULL GLASS OF ORANGE JUICE OR EAT A BANANA DAILY WHILE TAKING THIS MEDICATION
    4. TAKE WITH FOOD OR MILK

405. Which of the following is a side effect reported with phenytoin?
    1. slurred speech
    2. insomnia
    3. excessive thirst
    4. excessive sweating

406. Histamine H-2 receptor antagonists are used to treat which of the following conditions?
    1. asthma
    2. diabetes
    3. hypertension
    4. gastric ulcers

407. Which of the following answers best describes the primary indicated use for nizatidine?
    1. joint inflammation
    2. duodenal ulcer
    3. urinary burning
    4. constipation

408. Which of the following is a side effect reported with naproxen?
    1. muscle cramps
    2. numbness
    3. indigestion
    4. sore throat

409. Which is an appropriate auxiliary label to use with ketorolac?
    1. AVOID SMOKING WHILE TAKING THIS MEDICATION
    2. MAY CAUSE DROWSINESS OR DIZZINESS
    3. DO NOT CRUSH
    4. DISCARD UNUSED PORTION

410. Which is an appropriate auxiliary label to use with prazosin?
    1. TAKE AFTER MEALS
    2. MAY CAUSE DROWSINESS OR DIZZINESS
    3. TAKE ON AN EMPTY STOMACH
    4. AVOID PROLONGED EXPOSURE TO SUNLIGHT

411. Coadministration of diclofenac and which of the following may produce an unwanted result?
    1. hormones
    2. laxatives
    3. thiazides
    4. vasodilators

412. Which drug is not a platelet inhibitor?
    1. ticlopidine
    2. ranitidine

3. abciximab
4. aspirin

413. What is the brand name for omeprazole?
    1. PRECOSE
    2. OGEN
    3. PRILOSEC
    4. PRISCOLINE

414. What information should you give the patient taking ibuprofen?
    1. Report prolonged heartburn and abdominal pain.
    2. Double up on missed doses.
    3. Medication may cause drowsiness; therefore, test reaction to drug before operating a vehicle.
    4. answers 1 and 3

415. What information should you give the patient taking potassium chloride?
    1. Do not use salt substitutes while on this medication.
    2. Dissolve the liquid, powder, and effervescent tablet forms in water or juice, and drink slowly.
    3. There is no concern for wax coating from some tablets that may be found in stool.
    4. all of the above

416. Which of the following is the generic name for FOLEX?
    1. metaproterenol
    2. folic acid
    3. methotrexate
    4. flutamide

417. Which of the following drug classes or groups is represented by digoxin?
    1. antihistamine
    2. ACE inhibitor
    3. cardiovascular agent
    4. bronchodilator

418. Which of the following answers best describes the primary indicated use for methotrexate?
    1. leukemias
    2. psoriasis
    3. rheumatoid arthritis
    4. all of the above

419. Which of the following answers best describes a unique association with cefaclor?
    1. Do not freeze liquid preparation.
    2. Entire course of medication is 10–14 days.
    3. Take medication in equal intervals around the clock.
    4. all of the above

420. Coadministration of pravastatin and which of the following may result in a potentially serious or even life-threatening drug interaction?
    1. pseudoephedrine
    2. docusate sodium

3. gemfibrozil
4. acetaminophen

421. Which is an appropriate auxiliary label to use with ibuprofen?
    1. FINISH ALL MEDICATION UNLESS OTHERWISE DIRECTED BY PRESCRIBER
    2. CAUTION: FEDERAL LAW PROHIBITS THE TRANSFER OF THIS DRUG TO ANY PERSON OTHER THAN THE PATIENT FOR WHOM IT WAS PRESCRIBED
    3. TAKE WITH FOOD
    4. TAKE MEDICATION ½ HOUR BEFORE MEALS

422. What information should you give the patient taking lisinopril?
    1. Report yellowing of the skin.
    2. Loss of appetite while taking this medication is usual.
    3. Report persistent cough.
    4. Any dietary style is permissible while on this medication.

423. Coadministration of famotidine and which of the following may produce an unwanted result?
    1. itraconazole
    2. ketaconazole
    3. glycopyrrolate
    4. answers 1 and 2

424. Using the key word "nonsedating" as the basis for your decision, which of the following does not belong?
    1. loratidine
    2. terfenadine
    3. astemizole
    4. chlorpheniramine

425. Which of the following is a primary use for clonazepam, while generalized anxiety and panic disorder is the primary indication for the similarly sounding drug, clorazepate?
    1. obsessive-compulsive disorder
    2. peptic ulcer disease
    3. seizures
    4. inflammation

426. What is the available dosage form for paclitaxel?
    1. IM
    2. SQ
    3. IV
    4. PO

427. Which of the following is a side effect reported with ranitidine?
    1. rapid heart rate
    2. nausea
    3. malaise
    4. muscle cramps

428. Which of the following answers best describes a unique association with diltiazem?
    1. Medication causes metallic taste.
    2. Do not crush the sustained release form of the drug.

Drugs

3. Avoid antacids while taking this drug.
4. Medication causes urine to change color.

429. Which of the following answers best describes a unique association with digoxin?
   1. Chills indicate toxicity.
   2. Blood in stools indicates toxicity.
   3. Visual disturbances indicate toxicity.
   4. Loss of coordination indicates toxicity.

430. Which of the following is the generic name for DARVOCET-N?
   1. propafenone
   2. propoxyphene napsylate/acetaminophen
   3. propantheline
   4. dantrolene

431. Which of the following drug classes or groups is represented by amoxicillin?
   1. antibiotic agent (beta-lactams)
   2. antibiotic agent (tetracyclines)
   3. antibiotic agent (fluoroquinolones)
   4. antibiotic agent (sulfonamides)

432. Which of the following answers describes an indicated use for medroxyprogesterone?
   1. contraception
   2. endometrial carcinoma
   3. leukemia
   4. organ transplant

433. Which of the following drugs is most often prescribed QD.
   1. ampicillin capsules
   2. carisoprodol tablets
   3. digoxin tablets
   4. acetaminophen tablets

434. Which of the following answers best describes a unique association with enalapril?
   1. causes metallic taste
   2. turns urine orange
   3. Antacids reduce effectiveness.
   4. Discontinue use during second and third trimesters during pregnancy.

435. How would you classify or group potassium chloride?
   1. intravenous additive
   2. vitamin
   3. electrolyte
   4. trace element

436. Which of the following is a common side effect reported with glipzide use?
   1. musculoskeletal disturbances
   2. respiratory disturbances
   3. depression
   4. gastrointestinal disturbances

437. Coadministration of digoxin and which of the following may produce an unwanted result?
    1. potassium-sparing diuretics
    2. potassium-wasting diuretics
    3. analgenics
    4. cough suppressants

438. What accounts for the way amlodipine works?
    1. The drug prevents sodium overload.
    2. The drug dilates coronary vessels by relaxing the smooth muscles of the coronary vessels.
    3. The drug balances the elimination and reuptake of sodium, potassium, and calcium.
    4. The drug slows the pumping action of the heart.

439. Which of the following drug classes or groups is represented by conjugated estrogens?
    1. immunosuppressants
    2. blood modifiers
    3. hormones
    4. autonomic agents

440. Which of the following answers best describes the primary indicated use for phenytoin?
    1. seasonal affective disorder
    2. hypertension
    3. ulcers
    4. tonic-clonic (grand mal) seizures

441. Which of the following answers best describes a unique association with sertraline?
    1. Refrain from drinking alcohol while taking this drug.
    2. Clean container before instilling dose to each eye.
    3. take with a full glass of water
    4. causes hypotension

442. Which of the following is the generic name for CARDIZEM?
    1. atenolol
    2. verapamil
    3. diltiazem
    4. lisinopril

443. Which of the following drugs is a HMG-Co A reductase inhibitor?
    1. albuterol
    2. atenolol
    3. glyburide
    4. simvastatin

444. Which of the following is the trade name for ethinyl estradiol/levonorgestrel?
    1. OTHRO-NOVUM
    2. ORTHO-EST
    3. NORDETTE
    4. OGEN

Drugs

445. Which answer best describes the usual adult dose for sertraline?
    1. 25 mcg. qhs
    2. maximum 200 mg./day
    3. 20 mg./Kg BID
    4. up to 200 mcg./day
446. ____ is to alprazolam as ZANTAC is to XANAX.
    1. allopurinol
    2. alteplase
    3. ramipril
    4. ranitidine
447. Which of the following answers best describes a unique association with lovastatin?
    1. exempted from child proof containers
    2. low-salt diet
    3. take with evening meal
    4. take with plenty of fluid
448. Which of the following answers best describes the primary indicated use for propoxyphene napsylate/acetaminophen?
    1. mild to moderate pain
    2. inflammation
    3. fever
    4. muscle spasms
449. Which of the following is a common side effect reported with glyburide use?
    1. insomnia
    2. gastrointestinal disturbances
    3. hair loss
    4. depression
450. Which of the following is a side effect reported with potassium chloride?
    1. flatulence
    2. oral ulcers
    3. salty taste
    4. bruising
451. Which is an appropriate auxiliary label to use with glyburide?
    1. TAKE WITH FOOD OR MILK
    2. MAY CAUSE DROWSINESS
    3. DO NOT CRUSH
    4. DO NOT TAKE ASPIRIN
452. Which of the following answers best describes the primary indicated use for lisinopril?
    1. COPD
    2. hypertension
    3. CHF
    4. answers 2 and 3
453. What is the generic name for the drug AMOXIL?
    1. ampicillin
    2. amoxicillin

3. erythromycin
4. penicillin

454. Which of the following answers best describes a unique association with omeprazole?
    1. causes urine discoloration
    2. must be swallowed whole
    3. Take medication with plenty of fluid.
    4. Store in glass.

455. Which is an appropriate auxiliary label to use with fluoxetine?
    1. TAKE ON A FULL STOMACH
    2. MAY CAUSE DROWSINESS
    3. TAKE ON AN EMPTY STOMACH
    4. MAY CAUSE DISCOLORATION OF URINE OR FECES

456. Which of the following answers best describes the most common indication for ciprofloxacin?
    1. hepatitis
    2. infection caused by susceptible strains of organisms
    3. endocarditis
    4. viral infection of the upper respiratory tract

457. Which is a required statement to use with propoxyphene napsylate/acetaminophen either as a strip label or on the primary label?
    1. Caution: Federal law prohibits the transfer of this drug to any person other than the person for whom it was prescribed.
    2. Federal law prohibits dispensing without a prescription.
    3. Finish all this medication unless otherwise directed by prescriber.
    4. Do not use after (date).

458. Which of the following answers best describes the usual adult oral dosage for captopril?
    1. maximum dose: 450 mg. per day
    2. maximum dose: 50 mg. QID
    3. dose range: 50–200 mg.
    4. dose range: 25–100 mg.

459. Which of the following is a side effect reported with loratidine?
    1. joint pain
    2. dry mouth
    3. fever
    4. visual disturbances

460. What information should you give the patient taking hydroxychloroquine?
    1. Complete the full course of therapy.
    2. Protect eyes from sunlight.
    3. Report ringing in the ears (tinnitus).
    4. all of the above

461. Coadministration of glyburide and which of the following may produce an unwanted result?
    1. NSAIDs
    2. salicylates
    3. thiazides
    4. all of the above

462. What information should you give the patient taking pravastatin?
    1. Use in conjunction with dietary therapy.
    2. Take medication only after eating high fat meals.
    3. Take a blood test for lipid profile every 3 weeks.
    4. Double dose if a dose is missed.
463. Which of the following answers best describes a unique association with conjugated estrogens?
    1. Use of medication is restricted to osteoporosis in elderly patients.
    2. Avoid alcohol.
    3. Do not use during pregnancy.
    4. Chew tablets before swallowing.
464. How would you classify or group nifedipine?
    1. calcium channel blocker
    2. cardiovascular agent
    3. antiseizure drug
    4. answers 1 and 2
465. Paroxetine is to PAXIL as ____ is to TAXOL.
    1. pamidronate
    2. pancuronium
    3. pemoline
    4. paclitaxel
466. Coadministration of prazosin and which of the following may produce an unwanted result?
    1. verapamil
    2. codeine
    3. prednisone
    4. ergotamine
467. Which of the following is the trade name for paroxetine?
    1. PAVABID
    2. PROVERA
    3. PENTAM
    4. PAXIL
468. Which of the following drug classes or groups is represented by clarithromycin?
    1. anti-infectives (beta-lactams)
    2. anti-infectives (aminoglycosides)
    3. anti-infectives (cephalosporins)
    4. anti-infectives (macrolides)
469. What best describes a unique association with warfarin sodium?
    1. may cause a change in urine color
    2. should be taken with food
    3. causes a metallic taste
    4. double up missed doses
470. What information should you give a patient taking cyclophosphamide?
    1. Do not take the drug at nighttime.
    2. Urinate frequently.
    3. Drink fluids liberally before and after each dose.
    4. all of the above

471. Which of the following is the generic name for MINIPRESS?
    1. pravastatin
    2. minoxidil
    3. probucol
    4. prazosin
472. Which of the following is a side effect reported with lovastatin?
    1. excessive thirst
    2. flatus
    3. chills
    4. wheezing
473. Which of the following answers best describes a unique association with amoxicillin?
    1. Take drug with food to reduce gastric upset.
    2. Avoid antacids while taking drug.
    3. Take drug around the clock in equal intervals.
    4. Store drug in glass container.
474. What information should you give the patient taking loratidine?
    1. Drink fluid liberally.
    2. Reduce coffee consumption while taking this medication.
    3. Take medication with a full glass of orange juice.
    4. Double up on missed doses.
475. How would you classify or group pravastatin?
    1. antiulcer agent
    2. antidepressive agent
    3. antihypertensive agent
    4. antihyperlipidemic agent
476. Coadministration of loratidine and which of the following may produce an unwanted result?
    1. calcium channel blocking agents
    2. triazole antifungal agents
    3. laxatives
    4. antiemetic agents
477. The protease inhibitors are used primarily to treat which of the following conditions?
    1. HIV infection
    2. asthma
    3. diabetes
    4. hyperlipidemia
478. What is the trade or brand name for the generic drug cefaclor?
    1. CEFTIN
    2. CEFZIL
    3. CECLOR
    4. CEFOTAN
479. Which of the following answers best describes the primary indicated use for potassium chloride?
    1. hyponatremia
    2. hyperkalemia
    3. hypoglycemia
    4. hypokalemia

480. Which best describes the class of drugs to which ranitidine belongs?
    1. NSAIDs
    2. histamine H-2 antagonists
    3. ACE inhibitors
    4. steroids

481. Which of the following drug classes or groups is represented by diltiazem?
    1. analgesic agent (opioid)
    2. anti-infective agent (beta-lactam)
    3. cardiovascular agent (calcium channel blocker)
    4. hormone agent (anabolic steroid)

482. What is the generic name for MOTRIN?
    1. ibuprofen
    2. ketoprofen
    3. fenoprofen
    4. indomethacin

483. Which of the following is a side effect reported with pravastatin?
    1. yellowing skin
    2. abdominal cramps
    3. fatigue
    4. hair loss

484. FELDENE is to ____ as SELDANE is to terfenadine.
    1. famotidine
    2. cyclobenzaprine
    3. piroxicam
    4. sulindac

485. What is the brand name for phenytoin?
    1. CELONTIN
    2. ZARONTIN
    3. NEURONTIN
    4. DILANTIN

486. Which of the following is a side effect reported with medroxyprogesterone?
    1. bleeding
    2. changes in menstrual flow
    3. edema
    4. all of the above

487. Which answer best describes the usual adult dose for simvastatin?
    1. up to 40 mg. once a day
    2. maximum of 40 mcg. daily
    3. 40 mg. in the AM and PM
    4. 4 mg. qhs

488. BETOPTIC is to TIMOPTIC as betaxolol is to ____ .
    1. timolol
    2. tioconazole
    3. carteolol
    4. levobunolol

489. Which is an appropriate auxiliary label to use with furosemide?
    1. AVOID PROLONGED EXPOSURE TO SUNLIGHT
    2. FINISH ALL THIS MEDICATION UNLESS OTHERWISE DIRECTED BY PRESCRIBER
    3. TAKE ON AN EMPTY STOMACH
    4. DO NOT TAKE WITH DAIRY PRODUCTS, ANTACIDS, OR IRON PREPARATIONS

490. Which of the following answers best describes a unique association with ciprofloxacin?
    1. Avoid antacids containing magnesium or aluminum, or products with iron or zinc within 4 hours before or 2 hours after dosing.
    2. Store medication under refrigerated conditions.
    3. Do not crush or chew medication.
    4. Medication causes a metallic taste.

491. Which is an appropriate strip label to use with ranitidine?
    1. TAKE WITH FOOD
    2. PROTECT FROM LIGHT
    3. MAY CAUSE DROWSINESS
    4. MAY CAUSE DISCOLORATION OF URINE OR FECES

492. Coadministration of fluconazole and which of the folllowing may produce an unwanted result?
    1. metoprolol
    2. methocarbamol
    3. diphenhydramine
    4. warfarin

493. Which drug prepared parenterally should never be put in a PVC bag?
    1. Erythromycin
    2. Dopamine
    3. Ranitidine
    4. Paclitaxol

494. What is an appropriate strip label to use with cyclophosphamide?
    1. TAKE WITH FOOD OR MILK
    2. DRINK FLUID LIBERALLY
    3. DO TAKE DAIRY PRODUCTS OR ANTACIDS WITHIN 1 HOUR OF THIS MEDICATION
    4. MAY CAUSE DISCOLORATION OF URINE OR FECES

495. What information should you give the patient taking prazosin?
    1. Monitor blood pressure regularly.
    2. Maintain a low sodium diet.
    3. Rise slowly from a sitting or lying position.
    4. all of the above

496. What should you be sure to tell the patient taking AUGMENTIN?
    1. The medication may discolor the urine.
    2. The medication may cause drowsiness.
    3. Drink a daily glass of cranberry juice while taking this medication.
    4. Complete the full course of therapy.

Drugs

497. Which of the following best describes a unique association with nabumetone?
    1. take with an antacid
    2. take with or without food
    3. take TIDpc
    4. take PRN

498. Which of the following best describes the primary indicated use for levothyroxin?
    1. underactive thyroid
    2. prevention of pregnancy
    3. diabetes insipidus
    4. HIV infection

499. Coadministration of propoxyphene napsylate/acetaminophen and which of the following may result in a potentially serious or even life-threatening drug interaction?
    1. senna
    2. lactulose
    3. diphenoxylate
    4. carbamazapine

500. Which of the following is an antihistamine that is available in a nasal spray form?
    1. chlorpheniramine
    2. diphenhydramine
    3. azelastine
    4. loratidine

501. Which of the following answers best describes the usual oral dosage for medroxyprogesterone?
    1. 5 mg. daily
    2. 10 mg. BID
    3. dosing dependent on condition being treated
    4. none of the above

502. Which of the following is a side effect reported with methotrexate?
    1. edema
    2. oral ulcerations
    3. lacrimation
    4. belching

503. Of the listed answers, which one is a brand name for potassium chloride?
    1. KLONOPIN
    2. KAOCHLOR
    3. PLENDIL
    4. PLACIDYL

504. Which is the generic name for PRAVACHOL?
    1. prednisone
    2. probenecid
    3. pravastatin
    4. propafenone

505. Which of the following best describes a unique association with medroxyprogesterone?
    1. not recommended for use during first 4 months of pregnancy
    2. causes a metallic taste
    3. rinse mouth with water after each dose
    4. causes hallucinations

506. Coadministration of diltiazem and which of the following may produce an unwanted result?
    1. colony stimulating factors
    2. psychotropics
    3. muscle relaxants
    4. beta-blockers

507. Which of the following is the appropriate route of administration for a long-acting penicillin injection?
    1. IV
    2. IM
    3. SQ
    4. subclavian

508. How would you classify or group prazosin?
    1. ACE inhibitor/cardiovascular agent
    2. alpha adrenergic blocker/cardiovascular agent
    3. inotropic/cardiovascular agent
    4. calcium channel blocker/cardiovascular agent

509. Which label is appropriate to supplement the regular label on a prescription for conjugated estrogens?
    1. TAKE WITH FOOD
    2. TAKE ON AN EMPTY STOMACH
    3. MAY CAUSE DROWSINESS
    4. AVOID PROLONGED EXPOSURE TO SUNLIGHT

510. Diphenhydramine is to dimenhydrinate, as BENADRYL is to ___.
    1. DRAMAMINE
    2. DILANTIN
    3. DYAZIDE
    4. DONNATAL

511. Which of the following drug classes or groups is represented by enalapril?
    1. cardiovascular agent (beta blocker)
    2. cardiovascular agent (diuretic)
    3. cardiovascular agent (ACE inhibitor)
    4. cardiovascular agent (calcium channel blocker)

512. Which of the following answers best describes the primary indicated use for pravastatin?
    1. hyperkalemia
    2. hypernatremia
    3. hypercholesterolemia
    4. hypertension

Drugs

513. Which of the following answers best describes a unique association with captopril?
    1. Avoid use of drug during second and third trimesters of pregnancy.
    2. Take drug one hour before meals.
    3. Drug is found in the milk of nursing mothers.
    4. all of the above

514. Which of the following answers best describes the usual oral dosage for warfarin sodium?
    1. Therapeutic range is 10 mcg./day.
    2. Maintenance dose is between 2–10 mg./day.
    3. Loading dose is 20 mg..
    4. Dose depends on the weight of the patient.

515. Which is an appropriate auxiliary label to use with gold sodium thiomalate?
    1. AVOID PROLONGED EXPOSURE TO SUNLIGHT
    2. AVOID SMOKING WHILE TAKING THIS MEDICATION
    3. TAKE WITH FOOD OR MILK
    4. TAKE ON AN EMPTY STOMACH

516. Which answer best describes the usual adult dose for terfenadine?
    1. up to 120 mg./d
    2. 120 mg. BID
    3. 12.0 mcg./Kg., child or adult
    4. 120 mcg. TID

517. Coadministration of nabumetone and which of the following may result in a potentially serious or even life-threatening drug interaction?
    1. codeine
    2. insulin
    3. glipizide
    4. warfarin

518. What is the trade or brand name for the generic drug clarithromycin?
    1. ERYTHROCIN
    2. BIAXIN
    3. ZITHROMAX
    4. CLEOCIN

519. What special advice can you offer a patient taking aspirin?
    1. Take medication with food or milk to reduce gastric irritation.
    2. Take medication with a full glass of water to reduce gastric irritation.
    3. Finish all the medication unless the prescriber directed differently.
    4. Avoid exposure to sunlight.

520. Which of the following is a side effect reported with prazosin?
    1. arthralgia
    2. orthostatic hypotension

3. pruritis
4. urinary retention

521. Which of the following answers best describes a unique association with clarithromycin?
    1. Do not use in pregnant women.
    2. Do not refrigerate suspension.
    3. Do not use a mixed suspension after 14 days.
    4. all of the above

522. Which of the following drug classes or groups is represented by ciprofloxacin?
    1. anti-infectives (quinolones)
    2. cardiovasculars (beta blockers)
    3. psychotherapeutics (benzodiazepines)
    4. analgesics (opioids)

523. Which of the following answers best describes the most common indication for captopril?
    1. angina
    2. hypertension
    3. myocardial infarction
    4. tachycardia

524. What information should you give the patient taking fluoxetine?
    1. Avoid alcohol and alcohol-containing beverages.
    2. Determine the effect of the medication before operating a vehicle.
    3. A response may take several weeks.
    4. all of the above

525. Which is an appropriate auxiliary label to use with lisinopril?
    1. TAKE WITH FOOD
    2. MAY CAUSE DROWSINESS OR DIZZINESS
    3. DO NOT CRUSH. SWALLOW MEDICATION WHOLE
    4. AVOID EXPOSURE TO SUNLIGHT

526. Which of the following answers best describes a unique association with phenytoin?
    1. Do not crush medication.
    2. Medication causes hallucinations.
    3. Do not change brands.
    4. Swish oral suspension in mouth before swallowing.

527. Which is an appropriate auxiliary label to use with sertraline?
    1. DO NOT CHEW OR CRUSH
    2. PROTECT MEDICATION FROM LIGHT
    3. AVOID ALCOHOL
    4. AVOID PROLONGED EXPOSURE TO SUNLIGHT

528. Coadministration of levothyroxin and which of the following may produce an unwanted result?
    1. anticoagulants
    2. muscle relaxants
    3. analgesics
    4. laxatives

529. Cefuroxime can interact with which of the following drugs?
    1. psychotherapeutic drugs
    2. oral contraceptive drugs
    3. antiseizure drugs
    4. analgesic drugs
530. Of the following, what should the patient know about taking ranitidine?
    1. A 6 P.M. dosing may be more effective that a bedtime dose because acid secretion increases at around 7 P.M.
    2. Maintain a low salt diet for optimum effect.
    3. Store medication in a glass container.
    4. Do not crush or chew the medication.
531. What answer describes a unique association with conjugated estrogens?
    1. Increase vitamin B-6 and folic acid intake while taking this medication.
    2. Increase calcium intake while taking this medication.
    3. This medication renders iron supplements ineffective.
    4. Supplement medication with aspirin to reduce platelet aggregation.
532. What information should you give the patient taking propoxyphene napsylate/acetaminophen?
    1. Rise slowly from a sitting position.
    2. Cigarette smoking decreases effect of the medication.
    3. Avoid alcohol and sedatives while taking this medication.
    4. answers 2 and 3
533. Which of the following answers best describes a unique association with loratidine?
    1. This medication causes hallucinations.
    2. Drink plenty of water while taking this medication.
    3. Stay out of sunlight while taking this medication.
    4. This medication causes urine discoloration.
534. Which of the following answers best describes the most common indication for clarithromycin?
    1. infection caused by susceptible urinary tract pathogen
    2. infection of bony tissue
    3. infection caused by susceptible respiratory strains of pathogens
    4. infection of the gastrointestinal tract
535. Which of the following answers best describes the usual oral dosage for nifedipine?
    1. titrate adult dosages above 180 mg. per day
    2. 120 mg. twice a day
    3. adult range between 10–20 mg. per day
    4. adult range between 10–20 mg. three times a day
536. Coadministration of ethinyl estsradiol/levonorgestrel and which of the following may produce an unwanted result?
    1. codeine
    2. theophylline

3. guaifenesin
4. carisoprodol

537. Coadministration of ranitidine and which of the following may produce an unwanted reaction?
    1. cascara sagrada
    2. ketaconazole
    3. phenazopyridine
    4. sertraline

538. What is the brand name for azelastine?
    1. ZELAST
    2. LASTLONG
    3. ASTELIN
    4. HISTALINE

539. What information should you give the patient taking famotidine?
    1. Do not crush tablets.
    2. Once daily dosing should be taken preferably at bedtime.
    3. Stay out of the sunlight while taking this medication.
    4. Increase ascorbic acid intake while taking this medication.

540. Which of the following is the generic name for ORTHO-NOVUM?
    1. ethinyl estradiol/levonorgestrel
    2. mestranol/norethindrone
    3. ethinyl estradiol/norethindrone
    4. mestranol/norethynodrel

541. What is a usual sign of aspirin overdose?
    1. heart palpitations
    2. tinnitus
    3. diarrhea
    4. throbbing headache

542. THORAZINE is to chlorpromazine, as DIABINESE is to ____.
    1. chlorpheniramine
    2. chlorthalidone
    3. chlorprothixene
    4. chlorpropamide

543. Which of the following is a trade name for potassium chloride?
    1. K-TAB
    2. PENTASA
    3. PEPCID
    4. KLONOPIN

544. Which answer best describes the primary indicated use for fluconazole?
    1. bacterial infections
    2. candida infections
    3. viral infections
    4. tuberculosis

545. Hydroxyzine is to ____, ATARAX is to ATIVAN.
    1. lorazepam
    2. diazepam
    3. clonazepam
    4. flurazepam

Drugs

546. What auxiliary label would be appropriate for terfenadine?
    1. DRINK FLUIDS LIBERALLY
    2. MAY CAUSE DISCOLORATION OF URINE OR FECES
    3. TAKE ON AN EMPTY STOMACH
    4. TAKE WITH FOOD

547. What is a unique about the preparation and administration of paclitaxel?
    1. Take medication with plenty of fluid.
    2. Do not use with plasticized PVC equipment or devices.
    3. Rinse mouth after each dose.
    4. Protect drug from sunlight.

548. Which of the following is the generic name for KAY CIEL?
    1. sodium chloride
    2. potassium acetate
    3. calcium gluconate
    4. potassium chloride

549. Cephalexin may show a cross sensitivity to which of the following drugs?
    1. tetracycline
    2. erythromycin
    3. ciprofloxacin
    4. penicillin

550. ____ is to propranolol, as ISORDIL is to isosorbide.
    1. IMDUR
    2. IMODIUM
    3. INDERIDE
    4. INDERAL

551. HALCION is to triazolam, as RESTORIL is to ____.
    1. diazepam
    2. temazepam
    3. alprazolam
    4. lorazepam

552. Which is an appropriate strip label to use with colchicine?
    1. TAKE WITH FOOD OR MILK
    2. AVOID ALCOHOL
    3. DO NOT TAKE DAIRY PRODUCTS, ANTACIDS, OR IRON PREPARATIONS WITHIN 1 HOUR OF THIS MEDICATION
    4. MAY CAUSE DROWSINESS

553. Which best describes a characteristic for clonazepam?
    1. Schedule C-II
    2. Schedule C-III
    3. Schedule C-IV
    4. Schedule C-V

554. How would you classify propoxyphene napsylate/acetaminophen?
    1. diuretic
    2. blood modifier
    3. hormone
    4. analgesic

555. Which best describes the primary indicated use for famotidine?
    1. reduces gastric acid secretion
    2. reduces blood cholesterol
    3. reduces fluid retention
    4. stimulates gastric motility
556. Coadministration of cyclophosphamide and ____ may produce an unwanted result.
    1. allopurinol
    2. betamethasone
    3. carisoprodol
    4. diphenhydramine
557. Although cefaclor is absorbed better without food, why may the drug be taken with food?
    1. to reduce stomach upset
    2. to assure a timely schedule
    3. to clinically reduce the absorption rate
    4. to make the drug palatable
558. Coadministration of nizatidine and which of the following may produce an unwanted result?
    1. nifedipine
    2. ketoconazole
    3. acetazolamide
    4. ramipril
559. Which is an appropriate strip label to use with clonazepam?
    1. TAKE WITH PLENTY OF WATER
    2. TAKE ON AN EMPTY STOMACH
    3. TAKE WITH FOOD OR MILK
    4. AVOID ALCOHOL
560. Which of the following best describes the primary indicated use(s) for prazosin?
    1. hypertension and severe CHF
    2. angina and MI
    3. atherosclerosis
    4. hypercholesterolemia
561. What is the generic name for the drug DIABETA?
    1. metformin
    2. glyburide
    3. glipizide
    4. insulin
562. Glyburide belongs to which class of drugs?
    1. hypoglycemic agents
    2. hypertensive agents
    3. hypercholesterolemia agents
    4. hyperthyroid agents
563. How should cimetidine be taken?
    1. Take it with an antacid.
    2. Take it on an empty stomach.
    3. Take it as a supplement to a proton pump inhibitor.
    4. Take it with meals.

564. What would be helpful for a patient taking furosemide to know?
    1. Rise slowly from a lying or sitting position.
    2. Take medication at bedtime.
    3. Take medication early enough during the day to prevent nocturia.
    4. answers 1 and 3
565. What information should you provide a patient taking colchicine?
    1. Stop taking drug if GI symptoms occur.
    2. Take the drug with non-acidic juice.
    3. Continue to take the drug after symptoms have subsided.
    4. The 0.5 mg. and the 0.6 mg. strengths are interchangeable.
566. What is a side effect commonly reported with using propoxyphene napsylate/acetaminophen?
    1. sore throat
    2. bruising
    3. drowsiness
    4. dark urine
567. Which of the following answers best describes the usual oral dosage for amoxicillin?
    1. children: 100 mg. every 8 hours. adults: up to 500 mg. q8h
    2. children: 25–100 mg./Kg./d every 8 hours in divided doses. adults: 250–500 mg. every 8 hours
    3. children: 250 mg. BID. adults: 250 mg. QID
    4. children: 100 mg./Kg. every 8 hours. adults: 250 mg. per Kg. of body weight every 8 hours
568. What is the generic name for PROPULSID?
    1. famotidine
    2. lansoprazole
    3. cisapride
    4. omeprazole
569. What is the generic name for PEPCID?
    1. famotidine
    2. lansoprazole
    3. cisapride
    4. omeprazole
570. What is the generic name for PREVACID?
    1. famotidine
    2. lansoprazole
    3. cisapride
    4. omeprazole
571. What is the generic name for FELDENE?
    1. sulindac
    2. piroxicam
    3. cyclobenzaprine
    4. carisoprodol
572. What is the generic name for CLINORIL?
    1. sulindac
    2. piroxicam
    3. cyclobenzaprine
    4. carisoprodol

573. What is the generic name for FLEXERIL?
   1. sulindac
   2. piroxicam
   3. cyclobenzaprine
   4. carisoprodol

I. Assisting the Pharmacist in Serving Patients

Calculations

Basic Mathematics

Ratio
Proportion
Percentage

Pharmaceutical Calculations

Systems Conversions
Operational Uses

The calculations review section provides questions that cover basic mathematics and pharmaceutical calculations. Basic mathematics presents the reader with an opportunity to evaluate his or her strengths and weaknesses using basic mathematical operations, fundamental use of ratios and proportion, and an understanding of percentage. Pharmaceutical calculation questions deal with using the metric system, conversions, and the utilization of pharmaceutical calculations in practice.

1. Given the fraction 1/20, describe it in a percentage notation.
   1. 20%
   2. 1/2%
   3. 100%
   4. 5%
2. Display 15% in a decimal form.
   1. 1.5
   2. 0.15
   3. 15.0
   4. 0.015
3. What is the missing term in the proportion X:1 ml. :: 200 mEq.:50 ml.?
   1. 4 ml.
   2. 4 mMol.
   3. 4 mEq.
   4. 4 mcg.

Calculations

4. What is a valid fractional equivalent for 12.5%?
   1. 12.5/1000
   2. 125/1000
   3. 1.25/1000
   4. 0.125/1000
5. Add 22.8 mg. + 42.3 mg. + 37.6 mg. + 8.4 mg.
   1. 11.11
   2. 1.111
   3. 111.1
   4. 1111.0
6. Show ½ in a percentage notation.
   1. 2%
   2. 3%
   3. 50%
   4. ½%
7. Multiply 500.0 by 5.0123.
   1. 250.6150 (250.615)
   2. 2506.1500 (2506.15)
   3. 2560.15
   4. 2500.0
8. Which answer best describes the lowest fractional reduction for 25/100?
   1. ⅕
   2. ⅒
   3. 1/20
   4. ¼
9. Convert 12% to a decimal.
   1. 1200.0
   2. 0.12
   3. 1.2
   4. 12.0
10. What is the lowest reduced fraction for 15%?
    1. 15/100
    2. 0.15/100
    3. 3/20
    4. None of the above
11. What is the missing term in the proportion 2.4 mEq./7.2 mEq. = 4%/X?
    1. 12%
    2. 0.12 mEq.
    3. 0.12%
    4. 1.2%
12. Given the fraction ¾, which answer best identifies its decimal equivalent?
    1. 3.4
    2. 0.75
    3. 0.7
    4. 0.25

13. Convert 12.5% to a fraction in its lowest terms.
    1. 1/8
    2. 1/2
    3. 3/4
    4. 4/5
14. Display 12.5% in its decimal form.
    1. 1.25
    2. 12.5
    3. 0.125
    4. 125.0
15. Convert 1.7% to decimal.
    1. 170
    2. 0.17
    3. 0.017
    4. 1.7
16. Given the word expression *three quarters,* which best describes its lowest fractional reduction?
    1. 3/25
    2. 1/25
    3. 1/4
    4. 3/4
17. Convert 50% to a fraction in its lowest terms.
    1. 1/2
    2. 1/4
    3. 1/8
    4. 3/4
18. Given the fraction 1/20, convert this to a decimal notation.
    1. 5.5
    2. 1.2
    3. 0.05
    4. 0.5
19. Which is the missing term in the equation, 400 mcg./X = 40 mcg./1 ml.?
    1. 0.1 mcg.
    2. 10 ml.
    3. 0.1 ml.
    4. 10 mcg.
20. Given the fraction 1/2, convert it to a decimal notation.
    1. 0.5
    2. 0.2
    3. 0.02
    4. 0.05
21. Which is the missing term in the proportion, 2.4 mEq.: 5 ml. :: X : 1 ml.?
    1. 12%
    2. 0.48 ml.
    3. 0.48 mg.
    4. 0.48 mEq.

Calculations

22. Which answer best describes the lowest fractional reduction for the percentage 12.5%?
    1. 0.125/125
    2. 1/125
    3. 1/12.5
    4. 1/8
23. Multiply 4.738 by 0.
    1. 4.738
    2. 47.38
    3. 4.7380
    4. 0
24. What would a percent notation be for the expression *three quarters?*
    1. 25%
    2. 3%
    3. 75%
    4. 2.5%
25. Subtract 90% from 100%.
    1. 0.1
    2. 1.9
    3. 1.0
    4. 1.9
26. Which of the following answers best describes the fraction 25/100 verbally?
    1. twenty-five hundredths
    2. one-fifth
    3. twenty-five hundreds
    4. two and one-half
27. Divide 0.5 by 0.25.
    1. 2.0
    2. 0.2
    3. 20
    4. none of the above
28. Show 25/100 as a decimal.
    1. 0.025
    2. 0.25
    3. 2.5
    4. none of the above
29. Convert 1 ½ to a decimal.
    1. 3.2
    2. 15.0
    3. 0.15
    4. 1.5
30. Show the fraction ¼ as a percentage.
    1. 2.5%
    2. 25%
    3. 0.25%
    4. 14%

31. Which is the missing term in the proportion, 5000 units/1 ml. = 1250 units/X ml.?
    1. 0.5 ml.
    2. 4 ml.
    3. 0.25 ml.
    4. ¾ ml.
32. Given the word expression *three quarters,* which answer best identifies the fractional notation for this expression?
    1. ¾
    2. ⅜
    3. ¼
    4. ⅐
33. Convert 4/100 to a decimal.
    1. 0.4
    2. 0.04
    3. 0.004
    4. 4
34. Given the fraction 3/20, describe it in a percentage notation.
    1. 12%
    2. 15%
    3. 1.5%
    4. 0.15%
35. Which is the missing term in the proportion, 3:9 :: ____ :27?
    1. 9
    2. 12
    3. 14
    4. 6
36. Given the fraction ¼, which answer best identifies its decimal equivalent?
    1. 0.25
    2. 0.50
    3. 0.75
    4. 0.14
37. Add 0.056 Gm. + 0.0256 Gm. + 1.63 Gm.
    1. 1942 Gm.
    2. 194.2 mg.
    3. 19.42 Gm.
    4. 1,942 mg.
38. Given the word expression *three quarters,* which answer best identifies the decimal notation for this expression?
    1. 3.25
    2. 0.75
    3. 3
    4. 7.5
39. Which is the missing term in the expression, X : 1 ml. :: 400 mcg. : 10 ml. ?
    1. 4 ml.
    2. 400 mg.

Calculations

3. 4 mcg.
4. 0.04 mg.

40. Given the fraction 3/20, convert this to a decimal notation.
    1. 0.15
    2. 1.5
    3. 150.0
    4. 15.0

41. Divide 0.6875 by 0.8125.
    1. 8.462
    2. 8.4615
    3. 0.8462
    4. 84.6154

42. Which of the following best identifies the fractional notation for the expression *one-tenth*?
    1. 1/100
    2. 1/5
    3. 110
    4. 1/10

43. Subtract 12% from 87%.
    1. 0.99
    2. 7.5
    3. 0.75
    4. 0.075

44. *One hundred twenty-five thousandths* is expressed as a decimal by which of the following answers?
    1. 1.25
    2. 12.5
    3. 0.125
    4. 125.0

45. What term is missing in the expression, 40 mEq. : X :: 120 mEq. : 30 ml.?
    1. 0.333 mEq.
    2. 10 ml.
    3. 160 mEq.
    4. 16 ml.

46. Given the word expression *two-tenths*, which answer best describes the percent notation for this expression?
    1. 0.02%
    2. 0.2%
    3. 20%
    4. 2%

47. Convert 1.5 to a percent.
    1. 1.5%
    2. 15%
    3. 0.15%
    4. 150%

48. What is the lowest fractional reduction for the expression *fifteen-hundredths?*
    1. 15/100
    2. 1/7
    3. 1/8
    4. 3/20
49. Convert 0.5 to a percent.
    1. 50%
    2. 5%
    3. 0.05%
    4. 500%
50. Given the fraction 5/10, which answer identifies its decimal equivalent?
    1. 5.1
    2. 0.05
    3. 0.5
    4. 0.510
51. Subtract 23.75 ml. from 118 ml.
    1. −94.25 ml.
    2. 9.425 ml.
    3. −0.9425 ml.
    4. 94.25 ml.
52. Given the word expression *one-quarter,* which answer best identifies the fractional notation for this expression?
    1. 1/4
    2. 1/2
    3. 3/4
    4. 3/8
53. Convert 1.05 to a percent.
    1. 10.5%
    2. 1.05%
    3. 105%
    4. 1%
54. Given the fraction 75/100, which answer notes the percent equivalent?
    1. 75%
    2. 7.5%
    3. 0.75%
    4. 175%
55. What is the missing term in 234 mg./117 mg. = 23.4% /X?
    1. 0.117 mg.
    2. 11.7 mg.
    3. 11.7%
    4. 117 mcg.
56. What is the lowest fractional reduction for the fraction 5/100?
    1. 5/10
    2. 1/20
    3. 1/2
    4. 5/20

Calculations

57. Subtract 1.25 from 1.075.
    1. −0.175
    2. 0.175
    3. 1.750
    4. −1.750
58. Given the expression *one-tenth,* which answer best identifies the decimal notation for this expression?
    1. 0.01
    2. 0.001
    3. 0.1
    4. 1.0
59. Supply the missing term in the proportion 5 Gm./10 ml. = X/100 ml.
    1. 50 ml.
    2. 5%
    3. 50 Gm.
    4. 0.5 mg.
60. As a percent notation, how would you express one hundred twenty-five thousandths?
    1. 125%
    2. 1.25%
    3. 1250%
    4. 12.5%
61. Divide 0.6667 by 0.3333.
    1. 2.003
    2. 20.03
    3. 2000
    4. 2.0003
62. The expression *two-tenths* can be described in its lowest fractional reduction as which of the following?
    1. 1/5
    2. 1/50
    3. 2/100
    4. 2/1
63. Which is the missing term in the proportion, 5 mg. : 2 ml. :: 2 mg. : X?
    1. 0.8 ml.
    2. 8 ml.
    3. 5 ml.
    4. 5 mg.
64. Given the fraction 15/100, which answer identifies the decimal equivalent?
    1. 1.5
    2. 0.15
    3. 15.0
    4. 0.015
65. Subtract 12 fl. oz. from 24.8 fl. oz.
    1. 36.8 fl. oz.
    2. 1.28 fl. oz.

3. 128 fl. oz.
4. 12.8 fl. oz.

66. What is the percent equivalent for 5/10?
    1. 50%
    2. 10%
    3. 15%
    4. ½%

67. Convert 1 to a percent.
    1. 0.1%
    2. 0.01%
    3. 10%
    4. 100%

68. Given the percentage 25%, which answer best describes the lowest fractional reduction?
    1. 25/100
    2. 2/4
    3. ¼
    4. 0.25/100

69. Which is the missing term in the equation 50 mEq./2.5 ml. = X/4 ml.?
    1. 80 ml.
    2. 8.0 ml.
    3. 80 mEq.
    4. 80 mg.

70. Given the fraction 75/100, which answer identifies the decimal equivalent?
    1. 0.075
    2. 7.5
    3. 0,0075
    4. 0.75

71. Convert ¼ to a percent.
    1. 25%
    2. 0.25%
    3. 75%
    4. 2.5%

72. Which answer best describes 0.05 in a word expression?
    1. five-hundredths percent
    2. one-fifth
    3. five-hundredths
    4. one-half

73. Add 187.4 mg. + 13 Gm. + 276 mg.
    1. 13.4634 mg.
    2. 134.634 mg.
    3. 13.4634 Gm.
    4. 134.4634 Gm.

74. Given the decimal 0.05, which answer best shows the lowest fractional reduction?
    1. ⅕
    2. 1/100

Calculations

3. 5/20
4. 1/20

75. Convert 3/4 to a percent.
    1. 0.75%
    2. 7.5%
    3. 75%
    4. 0.075%

76. Given the word expression *one-tenth,* which answer best describes the percent notation for this expression?
    1. 1%
    2. 10%
    3. 100%
    4. 0.1%

77. Which term in the proportion 1/X = 8/16 is missing?
    1. 12.8
    2. 0.5
    3. 1/2
    4. 2

78. Which of the following is the lowest fractional reduction for the expression *one hundred twenty-five thousandths?*
    1. 125/1000
    2. 25/200
    3. 1/8
    4. 5/40

79. Add 1.374 + 11.246 + 247.3 + 0.11
    1. 260.030
    2. 2.603
    3. 26.03
    4. 2600.30

80. Given the fraction 2/10, which answer identifies its decimal equivalent?
    1. 2.0
    2. 0.2
    3. 0.02
    4. 2.1

81. To make the proportion proper, what should the missing term equal in the equation 5 Gm. : 10 ml. :: X : 100 ml.?
    1. 0.5%
    2. 50 Gm.
    3. 0.5 Gm.
    4. 50 ml.

82. What is the percent equivalent for the fraction 15/100?
    1. 15%
    2. 1500%
    3. 150%
    4. 0.15%

83. Add 127 ml. + 12.7 ml. + 1.25 ml. + 1 L.
    1. 1140.95 ml.
    2. 11.4095 ml.

3. 1.14095 L.
4. answers 1 and 3

84. Reduce 5/10 to the lowest fraction.
    1. 2/1
    2. 1/50
    3. 10/5
    4. 1/2

85. Divide 0.8333 by 0.50.
    1. 1.67
    2. 16.66
    3. 166.7
    4. 0.167

86. Which answer shows 25% as a valid fractional equivalent?
    1. 25/100
    2. 1/25
    3. 1/50
    4. All of the above

87. Subtract 543.1 mg. from 1000 mg.
    1. 456.9 mg.
    2. 45.69 mcg.
    3. 4.569 Kg.
    4. 456 Gm.

88. Describe the lowest fractional reduction for seventy-five hundredths.
    1. 1/2
    2. 1/4
    3. 3/4
    4. 1/8

89. Divide 0.65 by 0.450.
    1. 144
    2. 0.144
    3. 14.44
    4. 1.44

90. What is the decimal equivalent for 1/10?
    1. 1.0
    2. 1.1
    3. 0.1
    4. 0.01

91. Multiply 80 × 3.012.
    1. 2409.600
    2. 24.96
    3. 240.960
    4. 2410

92. Show the fractional notation for the expression *one hundred twenty-five thousandths*.
    1. 125,000
    2. 125/1000
    3. 125
    4. 12.5000

Calculations

93. Divide 23.75 by 10.2.
    1. 23.28
    2. 2.33
    3. 0.23
    4. 0.238

94. Given the fraction 2/10, which answer best notes the percent equivalent?
    1. 0.02%
    2. 0.2%
    3. 20%
    4. 2%

95. Multiply 10 × 0.1.
    1. 1.0
    2. 11
    3. 10.1
    4. 0.9

96. What is the lowest fractional reduction for the fraction 15/100?
    1. 4/20
    2. 3/20
    3. 1.5/100
    4. 15/10

97. Divide 0.45 by 0.650
    1. 6.923
    2. 0.69
    3. 6.9
    4. 69

98. How would you express the decimal, 0.5, in words?
    1. five-hundreds
    2. five-tenths
    3. five-hundredths
    4. five-percent

99. Which is the missing term for 3 mMol./50 ml. = 150 mMol./X?
    1. 2500 mMol.
    2. 2500 ml.
    3. 250 ml.
    4. 2.5 mMol.

100. Answer with the percent notation for the expression *twenty-five hundredths*.
    1. 2.5%
    2. 250%
    3. 25%
    4. 0.25%

101. Add 34.79502 + 24.1 + 14.1 + 10.1.
    1. 83.79502
    2. 837.9502
    3. 83.795
    4. 83.8

102. How would 75% be written in the lowest reduced fraction?
    1. ¾
    2. 75/100
    3. 15/20
    4. 6/8
103. Multiply 0.001 × 2.102.
    1. 0.2102
    2. 21.02
    3. 0.002
    4. 2.0012
104. Give a fractional meaning to the word expression *five-hundredths*.
    1. 5/100
    2. 100/5
    3. 5/10X100
    4. 500
105. Add 0.3 + 0.08 + 0.006 + 0.105.
    1. 491
    2. 4.91
    3. 0.491
    4. 49.1
106. The decimal 0.05 can be described in what percent notation?
    1. 0.5%
    2. 0.55%
    3. 5%
    4. 5.5%
107. Multiply 06.01 × 10.60.
    1. 63.71
    2. 16.61
    3. 1.76
    4. 67.31
108. The percentage 75% has which valid fractional equivalent?
    1. 75/100
    2. ¾
    3. 15/20
    4. all of the above
109. Multiply 12.75 × 2.15.
    1. 27.41
    2. 0.593
    3. 59.30
    4. 5.93
110. Which of the following is the decimal form for the percentage 25%?
    1. 2.5
    2. 0.25
    3. 0.025
    4. 25.0
111. Multiply 0.9 by 1.10.
    1. 1
    2. 0.909

Calculations 85

    3. 9.09
    4. 9.909

112. Given the word expression *five-tenths* which answer identifies the fractional notation for this expression?
    1. 510
    2. 5/10
    3. 1/4
    4. all of the above

113. Divide 5.75 by 0.375.
    1. 15.333
    2. 1.533
    3. 0.153
    4. 153.3

114. Which of the following is a word translation for the decimal 0.15?
    1. fifteen-tenths
    2. fifteen-thousandths
    3. fifteen-hundredths
    4. fifteen-hundreds

115. Which is the missing term in X : 100% :: 25 Gm. : 50 Gm.?
    1. 50%
    2. 0.5 Gm.
    3. 5 ml.
    4. 5%

116. Given the fraction 2/10, which answer best decsribes the lowest fractional reduction?
    1. 1/5
    2. 2/100
    3. 1/50
    4. none of the above

117. Multiply 1.07 × 2.01.
    1. 2.16
    2. 0.0215
    3. 0.216
    4. 2.15

118. Which of the following answers best describes the fraction 125/1000 in words?
    1. one hundred twenty-five thousand
    2. one hundred twenty-five thousandths
    3. one hundred twenty-five thousands
    4. eight

119. Convert 11.5% to a decimal.
    1. 1.15
    2. 11.5
    3. 115
    4. 0.115

120. What is the percent equivalent for the fraction 1/10?
    1. 1%
    2. 100%

3. 110%
4. 10%

121. Subtract 0.36 from 0.375.
    1. 0.150
    2. 1.150
    3. 0.015
    4. 15.0

122. What is the fractional equivalent for the decimal 0.75?
    1. 75/10
    2. 75/1000
    3. 75/100
    4. 75/10000

123. Add ⅜ + 0.943 + 11.21 + 27.
    1. 3.9528
    2. 39.528
    3. 395.28'
    4. 395.3

124. Which answer best shows the percent notation for the expression *one-quarter?*
    1. 4%
    2. 50%
    3. 25%
    4. 75%

125. Convert ½ to a percent.
    1. 0.5%
    2. 50%
    3. 5%
    4. 0.05%

126. Identify the decimal notation for the expression *five-tenths?*
    1. 0.510
    2. 5.10
    3. 0.5
    4. 5.5

127. Add 32.75 Gm. + 0.03629 Gm. + 0.00944 Gm.
    1. 32.79573 mg.
    2. 32.79573 Gm.
    3. 327.9573 Gm.
    4. 327957.3 Kg.

128. Given the decimal 0.15, which answer best shows the lowest fractional reduction?
    1. 15/100
    2. 15/10
    3. 3/2
    4. 3/20

129. Convert ⅛ to a percent.
    1. 1.25%
    2. 12.5%
    3. 125%
    4. 0.125%

Calculations

130. Of the following answers, which one denotes the fractional notation for the expression *two-tenths*?
    1. 1/10
    2. 5/10
    3. 210
    4. 2/10

131. Add 152.5 mg. + 42.5 mg. + 1 Gm. + 0.1 Gm.
    1. 1295.0 mg
    2. 1.295 Gm.
    3. 0.001295 Kg.
    4. all of the above

132. Which word expression best describes the decimal 0.1?
    1. one-hundredth
    2. one-tenth
    3. one-thousandth
    4. one

133. Supply the missing term in the proportion 40 mEq./20 ml. = 2 mEq./X.
    1. 0.5 mEq.
    2. 1 mEq.
    3. 1 ml.
    4. 0.5 ml.

134. Which of the following answers is a valid decimal equivalent for the fraction 125/1000?
    1. 0.125
    2. 1.25
    3. 12.5
    4. 0.0125

135. Convert 2/3 to a decimal.
    1. 0.67
    2. 6.7
    3. 0.066
    4. 0.067

136. How would you note the fractional equivalent for the decimal 0.05?
    1. 50/100
    2. 5/10
    3. 50/1000
    4. 5/100

137. Add 27.4 ml. + 135.5 ml. + 0.1 ml. + 1 ml.
    1. 164 ml.
    2. 163.01 ml.
    3. 0.164 L.
    4. answers 1 and 3

138. What is the lowest fractional reduction for the decimal 0.1?
    1. 1/10
    2. 1/100
    3. 10/50
    4. 1/1000

139. Divide 0.625 by 0.3125.
    1. 200
    2. 20
    3. 2.0
    4. 0.2
140. Given the fraction ¹²⁵⁄₁₀₀₀, what is the percent equivalent?
    1. 1.25%
    2. 12.5%
    3. 125%
    4. 0.125%
141. Subtract 5.5 cc. from 240 cc.
    1. 23.45 cc.
    2. 234.5 cc.
    3. 2.345 cc.
    4. 2.35 cc.
142. Describe the decimal 0.2 as a word expression.
    1. two-tens
    2. two-hundredths
    3. two-hundreds
    4. two-tenths
143. Convert 80% to a fraction in its lowest terms.
    1. ⅕
    2. ⅛
    3. ¾
    4. ⅘
144. What is another way the word expression *two-tenths* can be written?
    1. 2.0
    2. 0.2
    3. 0.02
    4. 0.002
145. Convert 0.125 to a percent.
    1. 0.125%
    2. 12.5%
    3. 1.25%
    4. 0.0125%
146. What is a fractional way of expressing a drug measured as fifteen-hundredths of a grain?
    1. ¹⁵⁄₁₀₀ gr.
    2. ¹⁵⁄₁₀₀₀ gr.
    3. ¹⁵⁄₁₀ gr.
    4. ¹⁵⁄₁ mg.
147. Subtract 0.75 Gm. from 3.1 Gm.
    1. −2.35
    2. 2.35
    3. 0.0235
    4. −0.235
148. What is the lowest reduced fraction for 0.25?
    1. ⅕
    2. ¼

Calculations

    3. 1/10
    4. 1/25

149. Convert 0.0075 to a percent.
    1. 7.5%
    2. 75%
    3. 0.75%
    4. 0.075%

150. Describe the decimal 0.75 in a percent notation.
    1. 75%
    2. 7.5%
    3. 0.75%
    4. 0.075%

151. Subtract 101.01 gr. from 1010.101 gr.
    1. 0 gr.
    2. 90.909 gr.
    3. 909.091 gr.
    4. 9.9091 gr.

152. What is another way of expressing the words *five-hundredths?*
    1. 0.5
    2. 0.05
    3. 0.005
    4. 0.0005

153. Which is the missing term in the proportion, X/25 Gm. = 4.06 mEq./0.5 Gm.?
    1. 0.1624 ml.
    2. 0.3248 mEq.
    3. 203 mEq.
    4. 20.3 mg.

154. What is 0.1 written as a percent?
    1. 1%
    2. 10%
    3. 100%
    4. 1000%

155. Subtract 0.35 from 35.035.
    1. 346.9
    2. 34.7
    3. 3.469
    4. 34.685

156. Subtract 0.35 mg. from 35.035 Gm.
    1. 35034.65 mg.
    2. 35.03465 Gm.
    3. 34.685 mg.
    4. answers 1 and 2

157. Multiply 3 mg. × 11.123 mg.
    1. 333.69 mg.
    2. 33.369 mg
    3. 33.639 Gm.
    4. 3.364 Gm.

158. Given a solution with a concentration of 20%, what decimal form would be needed to use in any calculation?
    1. 20.0
    2. 0.2
    3. 2.0
    4. 0.02

159. Supply the missing term in the equation 3mM : 45 mM :: 1 ml. : X.
    1. 15 ml.
    2. 0.07 ml.
    3. 1 ml.
    4. 1.5 ml.

160. A solution has a concentration of 0.15 written on the label. What percent is this?
    1. 1.5%
    2. 150%
    3. 15%
    4. 1500%

161. Convert 1/1000 to a decimal.
    1. 0.1
    2. 0.01
    3. 0.001
    4. 0.0001

162. Using the lowest fractional reduction, how many parts of ingredient A are there to ingredient B in a 0.5 preparation?
    1. 5/10
    2. 5/100
    3. 1/2
    4. 1/20

163. In order to complete this final mixture, how many units are needed in the proportion 10,000 u. : 1 ml. :: X u. : 0.5 ml.?
    1. 2500 u.
    2. 5000 u.
    3. 500 units.
    4. 50,000 units

164. Convert 1/8 to a decimal.
    1. 0.125
    2. 0.25
    3. 0.75
    4. 0.5

165. An unusual prescription calls for a seventy-five hundredths concentration of a product. How would this better be expressed as a percent?
    1. 75%
    2. 7.5%
    3. 0.075%
    4. 0.75%

Calculations

166. The order calls for 20 ml. of a preparation that you must make from an ampule of drug that contains 164 mg./ml. What amount of drug is needed to prepare 20 ml.?
    1. 3280 mg.
    2. 8.2 mg.
    3. 3.28 Gm.
    4. answers 1 and 3

167. A 5% solution contains what fraction equivalent?
    1. 1 part per 25 parts
    2. 5 parts per 10 parts
    3. 5 parts per 100 parts
    4. 50 parts per 1000 parts

168. Divide 0.1111 by 0.3333.
    1. 3.333
    2. 0.3333
    3. 30.33
    4. 3.30

169. What does a 0.1 concentration express in a fraction?
    1. 1/100
    2. 1/1000
    3. 1/10
    4. 1/1

170. Multiply 0.004 × 110.012.
    1. 0.440
    2. 4.40
    3. 4.048
    4. 440.05

171. An order reduces to 4 mEq. our standard need for 60 mEq. contained in a packaged 15 ml. product. How many milliliters are needed to provide 4 mEq.?
    1. 0.25 ml
    2. 1 ml.
    3. 0.5 ml.
    4. 1.25 ml.

172. In our 20% preparation, what is a valid fractional expression?
    1. 20/100
    2. 2/10
    3. 1/5
    4. all of the above

173. Convert 112% to a decimal.
    1. 1.12
    2. 11.2
    3. 112
    4. none of the above

174. Change 75% to a fraction in its lowest terms.
    1. 1/4
    2. 1/2

3. ¾
4. ⅛

175. In percent notation, what kind of solution is represented by 0.5?
    1. 50%
    2. 5%
    3. 0.5%
    4. 0.05%

176. Convert 0.05% to a decimal.
    1. 0.5
    2. 0.05
    3. 5.0
    4. 0.0005

177. Convert the decimal 0.75 to its lowest fractional form.
    1. ⅘
    2. 3/7
    3. ½
    4. ¾

178. A 10% solution has 1 part per how many parts of product?
    1. 20
    2. 1000
    3. 100
    4. 10

179. How many grams are needed to complete an order for a ⅕ preparation?
    1. 0.02 Gm.
    2. 0.002 Gm.
    3. 0.2 Gm.
    4. 0.5 Gm.

180. A 10% formulation provides how many grams of ingredient per 100 ml. of a product?
    1. 0.1 Gm.
    2. 0.01 Gm.
    3. 10 Gm.
    4. 1 Gm.

181. A prescription containing a total volume of 100 ml. has 5 cc. given for the first dose. What percent of the total volume is this amount?
    1. 20%
    2. 10%
    3. 5%
    4. 1%

182. How many 5 ml. doses are in 100 ml. of product?
    1. 5
    2. 10
    3. 15
    4. 20

183. If 75% of a 120 ml. prescription is taken by the patient, what volume is remaining?
    1. 25 ml.
    2. 75 ml.

3. 30 ml.
4. 90 ml.

184. If six teaspoonful doses have been given from a 4 fl. oz. prescription, what percent is this of the total prescription?
    1. 10%
    2. 25%
    3. 50%
    4. 75%

185. How is a precent volume-in-volume best defined?
    1. grams of solute in 100 milliliters Of solution
    2. milliliters of solute in 100 milliliters of solution
    3. 100 milliliters of solute in any fixed quantity of solution
    4. grams of solute in 100 grams of product

186. How many milliliters are required to complete a prescription for 7 days for MYLANTA 30 cc. pc and hs?
    1. 84mm ml.
    2. 630 ml.
    3. 840 ml.
    4. 84.0 ml.

187. How is a percent weight-in-weight best defined?
    1. grams of solute in 100 grams of solution
    2. grams of solute in 100 milliliters of solution
    3. volume of solute in 100 grams of solution
    4. grams of solute in a solution that equals any pre-calculated percentage

188. How is a percent weight-in-weight best defined
    1. 100 milliliters of solute in a finished product
    2. milliliters of solute in 100 milliliters of solution
    3. grams of solute in 100 milliliters of solution
    4. none of the above

189. You have magnesium oxide 400 mg. tablets on hand. On September 1, you receive a prescription that calls for 800 mg. qd for the first 3 days, 400 mg. BID for the next 2 days, and 400 mg. thereafter for the remainder of September. How many tablets do you need?
    1. 39
    2. 32
    3. 35
    4. 36

190. What is a ratio?
    1. the standard allowable percentage of error
    2. the relationship between two like quantities expressed as a common fraction
    3. a statistical tool
    4. a business management term referring to pricing practices

191. How many inches are in a meter?
    1. 39.37
    2. 2.54
    3. 36
    4. 15.432

192. Methylprednisolone IV has been increased to 80 mg. BID from 60 mg. BID. What new volume do you need from each vial containing 125 mg./5 cc. ?
    1. 2.4 cc.
    2. 3.2 cc.
    3. 0.8 cc.
    4. 5.0 cc.

193. How many centimeters are in an inch?
    1. 39.37
    2. 16.23
    3. 2.54
    4. 29.573

194. What is the abbreviation for centimeter?
    1. cent.
    2. deci.
    3. mm.
    4. cm.

195. A sodium polystyrene sulfonate po product is available as 15 Gm./30 ml. The product contains 20% sorbitol. How much sorbitol will the patient ingest with each 30 ml. dose?
    1. 60 mg.
    2. 6.0 ml.
    3. 0.6 Gm.
    4. 6.0 Gm.

196. What is the abbreviation for millimeter?
    1. mm.
    2. ml.
    3. milli.
    4. mil.

197. Which of the following is the abbreviation for kilogram?
    1. Kil.
    2. Kg.
    3. Klg.
    4. Km.

198. What is the abbreviation for milliliter?
    1. mm.
    2. cm.
    3. ml.
    4. Kl.

199. One ml. is to one cc. as _____ ml. is to one fluid drachm.
    1. 0.5
    2. 1
    3. 1/30
    4. 5

200. An order calls for the administration of 1000 mcg. of cyanocobalamine. How many milligrams will you dispense?
    1. 1 mg.
    2. 10 mg.

Calculations

      3. 100 mg.
      4. 0.1 mg.

201. One grain is equivalent to 0.065 Gm. = _____ mg.
    1. 65
    2. 14.432
    3. 480
    4. 437.5

202. What does the minim, fluid drachm, and fluid ounce each measure?
    1. gases
    2. fluid volume
    3. weight
    4. finished products

203. Potassium chloride for injection is available as 2 mEq./ml. in a 10 ml. single dose vial. Mixed in 100 cc. of NS to run over 1 hour, how long will it take to administer 10 mEq.?
    1. 10 minutes
    2. 15 minutes
    3. 30 minutes
    4. 60 minutes

204. What is a percentage?
    1. a computation based on 10
    2. a calculations using the log scale
    3. 100
    4. an expression that indicates the rate per hundred

205. How many milliliters of desmopressin injection are needed to fill an order requiring the patient to receive 0.2 mg. desmopressin per Kg. body weight when the patient weighs 150 pounds and the drug is available as 0.4 mcg. per ml.?
    1. 3.4 ml.
    2. 120 ml.
    3. 9.0 ml.
    4. 34 ml.

206. In household measure, how much medication should the patient take if the prescription reads, "Guaifenesin 15 cc. q4h po prn."?
    1. Take 1 teaspoonful orally every 4 hours as needed.
    2. Take 1½ teaspoonsful orall every 4 hours as needed.
    3. Take 1 tablespoonful by mouth every 4 hours as needed.
    4. Take 1 ounce by mouth every 4 hours as needed.

207. How many milligrams of desmopressin acetate injection is contained in one 10 ml. multiple-dose vial containing 4 mcg./ml.?
    1. 40 mg.
    2. 0.4 mg.
    3. 4 mg.
    4. 0.04 mg.

208. If a 1 milliliter ampule of phytonadione injection contains 10 mg. and the order calls for 7 mg. of phytonadione, what part of the ampule will contain the required amount?
    1. 1.4 ml.
    2. 0.7 ml.

3. 0.5 ml.
4. 0.35 ml.

209. How much guaifenesin would you dispense for 5 days supply if the prescription calls for "Guaifenesin 15 cc. q4h po prn."?
   1. 120 ml.
   2. 180 ml.
   3. 450 ml.
   4. 480 ml.

210. If 14 tablets of ciprofloxacin contain 7000 mg. of active ingredient, how much active ingredient is contained in 20 tablets?
   1. 10 Gm.
   2. 10000 mg.
   3. 0.01 Kg.
   4. all of the above

211. How many tablets are in a container holding 32500 mg. of active ingredient if each tablet contains 325 mg.?
   1. 100
   2. 50
   3. 25
   4. none of the above

212. How much hydrocortisone and menthol would you use in a compound for 240 ml. of an emollient lotion containing 2% hydrocortisone and ¼% menthol?
   1. HC @ 480 mg., menthol @ 600 mg.
   2. HC @ 4.8 Gm., menthol @ 600 mg.
   3. HC @ 480 mg., menthol @ 0.6 Gm.
   4. HC @ 48 mg., menthol @ 60 mg.

213. What is the cost for 24 mg. of an active ingredient used in a compound if the bulk bottle of the active ingredient costs $250 per gram?
   1. $9.00
   2. $6.00
   3. $3.00
   4. $1.50

214. If two tablets of acetaminophen contain 1000 mg. of active ingredient, how many tablets will make up the maximum daily allowable dosage of 4 Gm.?
   1. 8
   2. 4
   3. 2
   4. 1

215. How many teaspoonsful are in a tablespoonful?
   1. 2
   2. 3
   3. 4
   4. 5

216. For how many capsules should the prescriber have written on the following prescription: "Amoxicillin 250 mg., Sig: 1 TID for 10 days"?
    1. 10
    2. 20
    3. 30
    4. 40
217. Calculate the quantity of drug to dispense in a prescription that reads, "Guaifenesin, Sig: ii tsp. q4h around the clock for 3 days."
    1. 120 cc.
    2. 180 ml.
    3. 240 ml.
    4. 36 ml.
218. How many milliliters are in a standard teaspoonful?
    1. 3
    2. 5
    3. 10
    4. 15
219. *4* is to *IV* as *10* is to ____.
    1. L
    2. M
    3. X
    4. C
220. Nitroglycerin is available in a variety of strengths including 1/150 gr. What is this in milligrams?
    1. 6.6 mg.
    2. 15.0 mg.
    3. 0.15 mg.
    4. 0.4 mg.
221. How many liters are in a dozen pints of disinfectant?
    1. 5760 L.
    2. 5.76 L.
    3. 57.6 ml.
    4. 576 ml.
222. What is a proportion?
    1. a relationship between rates
    2. the amount of ingredients needed for compounding
    3. an expression of the equality of two ratios
    4. the quantity of active ingredients relative to the quantity of inert materials.
223. Add 2.5 Kg. + 125 mg. + 1000 mcg.
    1. 2500.126 Gm.
    2. 250.0126 mg.
    3. 2.500126 Kg.
    4. answers 1 and 3
224. Add 500 ml. + 1.3 L. + 0.25 L.
    1. 2050 L.
    2. 6.55 L.

3. 2.05 L.
4. 2.05 ml.

225. How many milliliters are there in a fluid ounce?
    1. 5
    2. 15
    3. 30
    4. 120

226. How many cc. are in 1 ml.?
    1. one
    2. five
    3. thirty
    4. one-hundred twenty

227. How many grams are in a milligram?
    1. 1000
    2. 0.001
    3. 100
    4. 0.01

228. What quantity on the prescription is represented by the Roman numeral "C"?
    1. ten
    2. twenty
    3. fifty
    4. one-hundred

229. How many grams are in a kilogram?
    1. 10
    2. 100
    3. ¹⁄₁₀₀₀
    4. 1000

230. What quantity on the prescription is represented by the Roman numeral "L"?
    1. five
    2. twenty
    3. fifty
    4. one hundred

231. How many cubic centimeters are in one milliliter?
    1. 1
    2. 10
    3. 100
    4. 1000

232. The Roman numeral "X" on a prescription indicates what quantity to dispense?
    1. five
    2. ten
    3. twenty
    4. fifty

233. What is the abbreviation for a cubic centimeter?
    1. cu.cen.
    2. cube

Calculations

3. C²
4. cc.

234. How do you read, "Nystatin s/s 5 cc. q6h."?
    1. Nystatin stable/suspension, 1 teaspoonful every 6 hours
    2. Nystatin, swish and swallow 1 teaspoonful every 6 hours
    3. Nystatin solubilized solution, 1 teaspoonful every 6 hours
    4. Nystatin, saliva susceptibility, 1 teaspoonful every 6 hours

235. What does the prefix "centi" mean?
    1. 1/10
    2. 1/100
    3. 1/1000
    4. 100 times

236. Which of the following answers is an equivalent of ½ ounce?
    1. one tablespoonful
    2. 15 cc.
    3. 15 ml.
    4. all of the above

237. What does the prefix "kilo" represent?
    1. 10 times
    2. 100 times
    3. 1000 times
    4. 1/1000 times

238. The difference between a dose of 1 "t." and 1 "T." on a prescription is the difference between ____ and ____, respectively.
    1. 15 ml. and 30 ml.
    2. titration and time (of the absorption of 1 dose)
    3. 5 cc. and 30 cc.
    4. 5 ml. and 15 ml.

239. What does the prefix "milli" mean?
    1. 1/10
    2. 1/100
    3. 1/1000
    4. 1000 times

240. Cubic centimeter is to cc. as ____ is to ml.
    1. milligram
    2. milliequivalent
    3. microgram
    4. milliliter

241. What is wrong with the prescription, "Give diphenhydramine po."?
    1. no strength
    2. no form
    3. no dosing instructions
    4. all of the above

242. What is the difference between 1 oz., 30 ml., 30 cc., and 2T?
    1. 2T is twice the amount of 1 oz.
    2. There is no difference.

3. 30 ml. and 30 cc. are equivalent, and 2T is double the amount of 1 oz.
4. It is an irrelevant answer because there is no logical relationship among the measures.

I. Assisting the Pharmacist in Serving Patients

Pharmacy & Medical Terminology

Abbreviations

Acronyms

Jargon

Symbols

1. A prescription for "hydrocortisone 1 percent ointment, apply TID" would most likely be written as which of the following?
   1. HC 1% oint. 3 Xid
   2. HC 1% ung. 3 id
   3. HC 1% oint. TID
   4. HC 1% oint. TOD
2. The acronym "CHF" refers to which of the following?
   1. congenital heart fibrillation
   2. conductive heart failure
   3. congestive heart failure
   4. cardiovascular hemostatic failure
3. The CDS refers to which of the following?
   1. cardiovascular drugs
   2. compounded drugs
   3. a governmental payment agency
   4. controlled drug substances
4. Most medical words best follow which of the following formulas?
   1. illness + body organ + symptom = word
   2. prefix + illness + itis = word
   3. affected body organ + symptom + syndrome = word
   4. prefix + Root + Suffix = word
5. Which of the following conditions may be managed by pharmaceutical management using a "step" therapy?
   1. hypertension
   2. chronic lower back pain
   3. obesity
   4. glaucoma
6. How would a prescriber write the directions for "zolpidem 5 mg. at bedtime if needed"?
   1. zolpidem 5 mg. qhs prn
   2. zolpidem 5 mg. qhs ad

3. zolpidem 5 mg. q 10 P.M.ad.lib
4. zolpidem 5 mg. qhsqs

7. Do ad. lib. and aq. dist. mean the same?
   1. yes
   2. no
   3. rarely
   4. frequently

8. What does the designation "SMZ" represent?
   1. sargramostim
   2. scopolamine
   3. sulfamethoxazole
   4. simvastatin

9. Is there a difference between a.l. and a.s.?
   1. yes
   2. no
   3. often
   4. rarely

10. The abbreviation a.l. seen on doctors' orders means ____?
    1. left ear
    2. prior to levigating
    3. left eye
    4. left arm

11. Seen often in compounding, ad. refers to ____.
    1. advance
    2. adhere
    3. before drying
    4. to, or up to

12. What does the designation "TMP" represent?
    1. triamterene
    2. trimethoprim
    3. trimeprazine
    4. triamcinolone

13. How would a prescription for "penicillin 500 milligram immediately, and 250 milligrams four times a day," most likely be written by the prescriber?
    1. penicillin 500 mg. at first, then 250 mg. Q.O.D.
    2. penicillin 500 mg. STAT, 250 mg. QID
    3. penicillin 500 mg. now, 250 mg. 4xid
    4. penicillin 500 mg. at once, then 250 mg. QD

14. Of right ear, left ear, both ears, or right eye, the abbreviation a.d. refers to which one?
    1. right eye
    2. left ear
    3. right ear
    4. both ears

15. What does a.c. stand for?
    1. before meals
    2. equal concentrations

3. before withholding
4. alternating concentrations

16. What is the common abbreviation $\overline{aa}$?
    1. before morning
    2. of each
    3. before meals
    4. each ear

17. How would the prescription for "sucralfate one gram four times a day before meals and at bedtime" be written?
    1. sucralfate 1 g. Q Xdachs
    2. sucralfate 1 gm. 4 idachs
    3. sucralfate T gm. QOD achs
    4. sucralfate T Gm. QID achs

18. What is the hospital category of drugs designated as PRN?
    1. patient restriction on narcotics
    2. medications which must be refrigerated because of a short expiration date
    3. This acronym refers to the method used to order the drug from the vendor.
    4. medications not routinely used on a specific schedule, but only when needed by the patient

19. What is the meaning of "dys" in the word dyspnea.
    1. difficult
    2. lacking
    3. slow
    4. very little

20. Hospital pharmacies may adopt for specific drugs an automatic stop-order policy usually designated by what acronym?
    1. ASOP
    2. STOP
    3. ASO
    4. AUTOSTOP

21. What are we referring to when we speak of a drug formulary?
    1. a listing of drugs selected by the hospital PT committee for inclusion in a pharmacy's inventory of drugs
    2. a directory of formulas that make up each drug
    3. a compounding directory
    4. a list of forms used to order various types of drugs such as controlled substances and alcohol

22. The root "vaso" in vasoconstrictor can denote drugs that affect what part of the anatomy?
    1. heart
    2. blood cells
    3. blood vessels
    4. lymph nodes

23. To what does the DEA refer?
    1. date of Exclusion Application
    2. drug Enforcement Administration

3. drug Entry Alert
4. drug Evaluation Agency

24. What is ambulatory care?
    1. the care provided to a patient in an ambulance
    2. the care provided to only walking patients
    3. health services provided on an outpatient basis
    4. this care refers to patients in a hospital who are able to walk

25. What would a prescription for "naphazoline eye drops, two drops in each eye twice a day when needed for redness" look like?
    1. naphazoline oph, 2 gtts OS BD prn redness
    2. naphazoline oph, 2 gtts OU BID prn redness
    3. naphazoline Otic, 2 gtts OU BID prn redness
    4. naphazoline Otic 2 gtts od BD prn redness

26. What is the meaning of "glyco" in the word glycosuria?
    1. sour
    2. salty
    3. sweetness
    4. tartness

27. It is important to read q.i.d. and q.o.d. directions carefully because
    1. q.i.d. is four times a day, q.o.d. is every other day.
    2. q.i.d. is every evening, q.o.d. is every day.
    3. q.i.d. is three times a day, q.o.d. is four times a day.
    4. q.i.d. is four times a day, q.o.d. is day or night.

28. What is bacteremia?
    1. a treatment for bacterial infection
    2. bacterial infection of the blood
    3. bacterial invasion of the intestine
    4. diarrhea

29. How many *qh* doses can be given during a *qd* schedule?
    1. 12
    2. 4
    3. 24
    4. 3

30. Rhinitis is usually a seasonal ailment. What part of the body is affected?
    1. nose
    2. mouth
    3. eyes
    4. ears

31. How often is a drug ordered p.r.n. given?
    1. once in the morning
    2. not to exceed 4 times a day
    3. no less than daily
    4. as needed

32. Drugs used to treat dermatitis are treating what condition?
    1. nasal polyps
    2. ringing in the ears

3. inflammed eyes
4. inflammation of the skin

33. The difference between the p.o. and p.r. routes of administration is:
    1. p.o. is by mouth, p.r. is rectally
    2. p.o. is rectally, p.r. is by mouth
    3. p.o. is parenterally, p.r. is enterally
    4. p.o. is topically, p.r. is by mouth

34. What part of the anatomy is effected by nephrotoxicity?
    1. the liver
    2. the heart
    3. the spleen
    4. the kidneys

35. In the sig., "Take one qhs," hs means
    1. at noon
    2. when needed
    3. at bedtime
    4. with water

36. The label should tell the patient to take a q3h w.a. prescription
    1. every 3 hours with antibiotics
    2. every 3 hours with water
    3. every 3 hours while awake
    4. every 3 hours as needed

37. What is the meaning of "per" in percutaneous?
    1. puncture
    2. soft
    3. pliable
    4. section of

38. By which route is a supp. given?
    1. p.o.
    2. p.r.
    3. q.d.
    4. s.a.

39. What is the meaning of "meta" in metabolism?
    1. minute
    2. false
    3. transformation
    4. deficient

40. STAT or stat on the prescription order means?
    1. statistically
    2. saturday dose
    3. now, immediately, or at once
    4. list the name of the drug

41. What is the meaning of "hypo" in hyponatremia?
    1. deficiency
    2. excessive
    3. dangerous
    4. drug interaction

Pharmacy & Medical Terminology

42. Of the following, what does the acronym "IV" mean?
    1. intravenous
    2. induce ventilation
    3. infectious virus
    4. inflammed veins
43. The acronym "IM" is found on the patient's orders. What does it mean?
    1. increase mobility
    2. intestinal mucosa
    3. inflammed muscle
    4. intramuscular
44. How would you describe the urine output using the meaning of "poly" in polyuria?
    1. deficient
    2. sweet odor
    3. excessive
    4. burning
45. In the following order, what do the abbreviations b.i.d., t.i.d., and q.i.d. mean?
    1. four times daily, three times daily, twice daily
    2. twice a day, three times a day, four times a day
    3. two times a day, four times daily, three times a day
    4. they all mean the same
46. What does the prefix "sub" mean in the word subcutaneous?
    1. beneath
    2. above
    3. irritated
    4. inflammed
47. Which abbreviation does not belong in the following group?
    1. tab.
    2. syr.
    3. susp.
    4. subq.
48. The ʒ symbol refers to?
    1. tablespoonful
    2. one-half tablespoonful
    3. teaspoonful
    4. 15 milliliters
49. "Angi" in angioplasty tells us that this procedure relates to what part of the anatomy?
    1. heart
    2. liver
    3. blood vessels
    4. kidneys
50. Which of the following does not belong?
    1. PZI
    2. PCN
    3. TCN
    4. gent

51. The "gingi" in gingivitis refers to an inflammation of what part of the anatomy?
    1. gums
    2. teeth
    3. throat
    4. tongue
52. "Take as directed" may be written as which of the following?
    1. ut. dict.
    2. u.d.
    3. ud
    4. all of the above
53. Which of the following best describes the root part "cyte" in the word hematocyte?
    1. cell
    2. color
    3. thickness
    4. flowability
54. What does the acronym "g/d" mean?
    1. grams per dose
    2. gastrointestinal disorder
    3. grams per day
    4. granuloma dissection
55. What information is provided by the NDC numbering system?
    1. the manufacturer
    2. the product name, strength, and dosage form
    3. the packaging size
    4. all of the above
56. The integumentary system includes words such as dermatitis. What does the common root "derma" mean?
    1. skin
    2. nails
    3. hair
    4. breasts
57. Which of the following is/are some designations used by manufacturers to indicate extended or long-acting dosage forms?
    1. SR
    2. SA
    3. CD
    4. all of the above
58. The "nephr" refers to what part of the genito-urinary system?
    1. liver
    2. spleen
    3. kidney
    4. bladder
59. What does the acronym "GI" mean?
    1. gastric infusion
    2. gastrointestinal

Pharmacy & Medical Terminology

3. gallbladder inflammation
4. gum infection

60. Which of the following answers best describes the number to be dispensed?
    1. DTD
    2. disp
    3. #
    4. all of the above

61. The common root "osteo" refers to what part of the musculo-skeletal system?
    1. bone
    2. tendon
    3. skull
    4. muscle

62. Which of the following acronyms or abbreviations on the prescription alerts us to the directions?
    1. AQ. DIST
    2. DTD
    3. SIG.
    4. STAT

63. The word arthritis can easily be dissected into "arthr" and "itis." Knowing that "itis" refers to an inflammation, which part of the skeletal system is described by "arthr"?
    1. tendons
    2. cartilage
    3. joint
    4. muscle

64. What does the acronym "5-FU" mean?
    1. 5 time fluid uptake
    2. 5th followup
    3. 5 pm followup
    4. 5 fluorouracil

65. Both "pneumo" and which other common root refer to lungs?
    1. broncho-
    2. laryn-
    3. pulmo-
    4. rhin-

66. What do the IM, IV, and SC routes of administration have in common about how drugs are administered?
    1. parenterally
    2. orally
    3. in larger volumes
    4. as admixtures

67. How is the label for "Instill two gtts. ou BID" typed out for the patient?
    1. instill 2 ophthalmic disks in both eyes twice a day.
    2. instill 2 drops in both eyes twice a day.

3. instill 2 drops in each ear twice a day.
4. instill 2 drops in the left ear twice a day.

68. The ear is to hearing as the root for the eye is to the root for seeing in which answer?
    1. oculo : opia
    2. auri : audit
    3. bleph : phaco
    4. ot : phaco

69. The acronym "ACE" is found on the patient's orders. What does it mean?
    1. angiotensin converting enzyme
    2. abdominal extension
    3. atrophic cardiac exacerbation
    4. arterial and cardiac evaluation

70. In the word intramuscular, what does the part "intra" mean?
    1. connected
    2. within
    3. striated
    4. smooth

71. The designation "ad" means "up to" or ____?
    1. left ear
    2. right nostril
    3. right eye
    4. right ear

72. In the word hypercalcemia, how do you interpret the part "hyper"?
    1. deficient
    2. precipitating
    3. noticeable
    4. excess

73. Ophthalmic is to eye as ____ is to liver.
    1. renal
    2. hepatic
    3. digestive
    4. pulmonary

74. Otic is to ear as renal is to ____ .
    1. liver
    2. eye
    3. kidney
    4. heart

75. In the word oliguria, what does the part "olig" mean?
    1. burning
    2. excessive
    3. odorous
    4. very little

76. The acronym "UTI" is found on the patient's orders. What does it mean?
    1. intrathecal infusion
    2. urinary tract infection

Pharmacy & Medical Terminology

3. upper thoracic infection
4. undiagnosed thyroid infection

77. A condition is noted by what suffix?
    1. ectasis
    2. itis
    3. pathy
    4. malacia

78. How is a dose taken s.l. administered?
    1. under the tongue
    2. in the cheek cavity
    3. sublingually
    4. answers 1 and 3

79. The acronym "GERD" is found on the patient's orders. What does it mean?
    1. gastroesophageal reflux disease
    2. geriatric evaluation for refractory dementia
    3. general evaluation by Registered Dietician
    4. gait evaluation required daily

80. What is the meaning of "cata" in catabolism?
    1. build up
    2. transformation
    3. downward
    4. minute

81. DPH is an acronym commonly used for which of the following drugs?
    1. dimenhydrinate
    2. diphenhydramine
    3. phenytoin
    4. diphenoxylate

82. What is the meaning of "anti" in antibiotic?
    1. oppose
    2. resistance
    3. tolerance
    4. none of the above

83. What does the acronym "DNR" mean?
    1. diagnosis not referenced
    2. do not resusitate
    3. does not respond
    4. day and night rotation

84. In the word erythrocyte, what does the part "erythro-" mean?
    1. deficient
    2. redness
    3. abundant
    4. pallor

85. The acronym "DNI" is found on the patient's orders. What does it mean?
    1. diagnosis not included
    2. do not infuse

3. do not intubate
4. discontinue new instructions

86. In the word contraceptive, what does the part "contra" mean?
    1. synergistic
    2. compliant
    3. enhancing
    4. against
87. What does the acronym "MDI" represent?
    1. medical diagnosis inquiry
    2. measured-drug inhaler
    3. physician internist
    4. metered-dose inhaler
88. What does SO_4 represent?
    1. sulfisoxazole
    2. sulfate
    3. sulfinpyrazone
    4. sulfasalazine
89. The acronym "DNA" is found on the patient's orders. What does it mean?
    1. deoxyribonucleic acid
    2. do not activate
    3. diphenhydramine
    4. do not aspirate
90. What does PO_4 represent?
    1. phosphate
    2. oxygenated polymer
    3. propofol
    4. propranolol
91. What does the doctor mean when he or she writes ut.dict. on the prescription?
    1. as directed
    2. Call his/her office without the patient's knowledge.
    3. Ask the patient to repeat the directions.
    4. Select the directions from the drug literature.
92. What does the acronym "COPD" mean?
    1. complete official position description
    2. continue observation of patient depression
    3. cancel out peripheral dilation
    4. chronic obstructive pulmonary disease
93. What does Cl represent?
    1. choline
    2. clindamycin
    3. clotrimazole
    4. chloride
94. In the word bradycardia, what does the part "brady" mean?
    1. accelerated
    2. slow
    3. non-reversible
    4. deficient

Pharmacy & Medical Terminology

95. The acronym "CNS" is found on the patient's orders. What does it mean?
    1. continue normal saline
    2. clear nasal sinuses
    3. central nervous system
    4. cancel neonatal screening

96. The common root "neuro-" refers to which part of the anatomy?
    1. nerve
    2. mind
    3. spinal cord
    4. brain

97. What does HCO_3 represent?
    1. hemophilus b conjugate vaccine
    2. bismuth
    3. bicarbonate
    4. hetastarch

98. What does Mg. represent?
    1. magnesium hydroxide
    2. magnesium
    3. magnesium sulfate
    4. manganese

99. What does the acronym "CHF" mean?
    1. caution hold fluids
    2. congestive heart failure
    3. continue hourly feedings
    4. complete history file

100. What does Ca represent?
    1. calcium
    2. calcitonin
    3. calcitriol
    4. calcium carbunale

101. The suffices "-algia" and "-dynia" refer to which type of symptom?
    1. fever
    2. headache
    3. pain
    4. constipation

102. The acronym "mg/d" is found on the patient's orders. What does it mean?
    1. magnesium per dose
    2. milligrams per day
    3. milligrams per dose
    4. magnesium in diet

103. The suffix "-oma" in a word often denotes which of the following?
    1. a softening
    2. a tumor
    3. a condition
    4. an inflammation

104. What does Na represent?
    1. naproxen
    2. saline
    3. sodium chloride
    4. sodium
105. What does the acronym "mg/kg/dose" mean?
    1. magnesium per kilogram dose
    2. manganese per kilogram of body weight per dose
    3. milligrams per kilogram of body weight per dose
    4. milligrams per kilogram of drug per dose
106. The suffix "-malacia" is most closely associated with which of the following?
    1. swelling
    2. malnutrition
    3. flowing
    4. softening
107. What does K represent?
    1. potassium phosphate
    2. potassium gluconate
    3. potassium
    4. ketaconazole
108. The acronym "NSAIDs" is found on the patient's orders. What does it mean?
    1. non-stable AIDs patient
    2. no susceptibility to AIDs
    3. normal saline for AIDs patients
    4. nonsteroidal anti-inflammatory drugs
109. What does Li represent?
    1. liothyronine
    2. lithium
    3. liotrix
    4. lisinopril
110. The suffix for inflammation is described by which of the following?
    1. cele
    2. itis
    3. mania
    4. rrhea
111. What does the acronym "OTC" mean?
    1. over-the-counter
    2. observed technical count
    3. ophthalmic treatment considered
    4. outside the curriculum
112. What does NH_4 represent?
    1. ammonium
    2. hydrogenated nitrogen
    3. nitric hydride
    4. heavy water

Pharmacy & Medical Terminology

113. The "rhin" in rhinitis describes this inflammation to be associated with what part of the respiratory system?
    1. windpipe
    2. lungs
    3. larynx
    4. nose

114. The acronym "/d" is found on the patient's orders. What does it mean?
    1. missing dose
    2. per day
    3. cut the dosage
    4. early discharge

115. What does CO_3 represent?
    1. heavy water
    2. carboxylate
    3. radioactive carbon
    4. carbonate

116. As "psych" refers to the mind, "cere" refers to which part of the nervous system?
    1. the head
    2. brain
    3. nerve
    4. vertebrae

117. What does the acronym "PCN" mean?
    1. post care nutrition
    2. platelet count plus neutrophils
    3. penicillin
    4. penicillamine

118. Although not a chemical formula, what does "LAC" or "lac" represent?
    1. lactulose
    2. leukotriene
    3. lactate
    4. left atrial compensation

119. What part of the musculo-skeletal system is affected by pain when referring to chondrodynia?
    1. joints
    2. skull
    3. tendons
    4. cartilage

120. The acronym "PUD" is found on the patient's orders. What does it mean?
    1. provide unit dose
    2. plicamycin-Uracil-Dactinomycin regimen
    3. pathogenic ureter disorder
    4. peptic ulcer disease

121. What is the difference between homeostasis and hemostasis?
    1. no difference
    2. Homeostasis refers to a balance in the body's physiologic systems, and hemostasis refers to properties of vasoconstriction and blood coagulation.
    3. Hemostasis refers to a balance in the body's physiologic systems, and homeostasis refers to properties of vasoconstriction and blood coagulation.
    4. Hemostasis refers to blood stability and homeostasis to blood coagulation.
122. Knowing that "dynia" refers to pain, drugs used to alleviate mastodynia affect which part of the anatomy?
    1. jaw
    2. breast
    3. neck
    4. mastoid bone
123. Although not the chemical formula, what does Ac or Acet represent?
    1. acetaminophen
    2. acetate
    3. gold
    4. acidophyllus
124. What does the acronym "RA" mean?
    1. rheumatoid arthritis
    2. right atrium
    3. regional ascites
    4. refractory anemia
125. Some drug side effects may include glossitis. The "gloss" refers to what part of the anatomy?
    1. vocal cords
    2. tongue
    3. cheeks
    4. salivary glands
126. What most commonly follows the "#" symbol on a prescription?
    1. a quantity
    2. a time
    3. a schedule
    4. a weight
127. Drugs that have a renal impact have an effect on which organ of the body?
    1. pancreas
    2. liver
    3. lungs
    4. kidney
128. Although gluc is not the chemical formula, written on an order, what does it represent?
    1. gluconate
    2. glucose
    3. glycuronic acid
    4. glucosamine

Pharmacy & Medical Terminology

129. The acronym "SOB" is found on the patient's orders. What does it mean?
    1. standard operational billing
    2. shortness of breath
    3. signs of bacteremia
    4. start observing breathing

130. "Emia" in hyperemia and "hem" in hemapoiesis refers to what part of anatomy?
    1. spleen
    2. blood
    3. veins
    4. arteries

131. Which of the following is a designation you will not find on a prescription order?
    1. #
    2. Disp.
    3. Sig:
    4. NB

132. What does a quantity followed by a "#" represent?
    1. a time in minutes
    2. a schedule while awake
    3. a weight in pounds
    4. a quantity allowing a 5% error

133. The "gastr" in gastritis refers to what part of the anatomy?
    1. liver
    2. kidney
    3. intestines
    4. stomach

134. What does the acronym "SSRI" mean?
    1. Standard System of Regional Infusion
    2. subsequent signs of respiratory inhibition
    3. selective serotonin reuptake inhibitor
    4. saturated solution of riboflavin injection

135. DOSS or DSS is used as what type of medication?
    1. laxative agent
    2. diabetic hypoglycemic agent
    3. antiemetic drug
    4. antidiarrheal drug

136. The root "cardium" in pericardium refers to what part of the anatomy?
    1. lungs
    2. aortic artery
    3. juglar vein
    4. heart

137. Diuretics affect the kidneys as cardiac drugs affect _____ ?
    1. the liver
    2. the heart
    3. the immune system
    4. the gallbladder

138. By reading the word tachycardia, what do we know from the prefix "tachy" about the beating of the heart?
    1. slowed
    2. accelerated
    3. irregular
    4. lacking
139. The acronym "TCAs" is found on the patient's orders. What does it mean?
    1. thoracic congested areas
    2. tetracycline agents
    3. tertiary catheter administration
    4. tricyclic antidepressants
140. The abbreviation "$\overline{ss}$" may refer to a saturated solution, but when associated with an amount or quantity, it means
    1. standardized substance
    2. selective substitution
    3. one-half
    4. ounce
141. The prefix "semi" gives what meaning to the word semipermeable?
    1. total
    2. block
    3. unmeasurable
    4. half
142. Which dosage form would be the best choice for administration of a drug through an NG tube?
    1. sol
    2. lot
    3. supp
    4. ung
143. In the word pericarditis, what best describes the closest meaning for the prefix "peri?"
    1. within
    2. around
    3. through
    4. of close proximity
144. What does the acronym "TCN" mean?
    1. tetracycline
    2. total care nutrition
    3. tetracaine
    4. ticarcillin
145. What could we deduce from a report that notes glycosuria?
    1. There is excessive urine output.
    2. Sugar is found in the urine.
    3. Urine output has diminished.
    4. Stools show blood content.
146. The Gm. in 100 Gm. of powder means
    1. grams
    2. grains

3. drops
4. none of the above

147. A person suffering from dyspnea shows what problem?
    1. difficulty in breathing
    2. snoring
    3. wheezing
    4. heart attack

148. The difference between C and c in a prescription is
    1. C means carefully, c means with.
    2. C means 100, c means caution.
    3. C means compound, c means capsules.
    4. C means 100, c means with.

149. Where is the bleeding occurring when located intracranially?
    1. around the heart
    2. within the eye
    3. around the spleen
    4. within the brain

150. A.M. could easily be taken for anterior muscle, but really means what?
    1. morning
    2. before meals
    3. after medicine
    4. evening

151. What is the meaning of the acronym "TFN"?
    1. the first nutrient
    2. take for now
    3. terminate forced nutrition
    4. til further notice

152. What does b.i.d. mean in the directions, "Take 1 tablet b.i.d."?
    1. before ingesting diet
    2. before bedtime
    3. twice a day
    4. none of the above

153. A physician's order which contains the acronym "HTFN" means what?
    1. have the family notified
    2. hold the first nutrient
    3. hold til further notice
    4. hydralazine-terazosin-felodipine-nadolol hypertensive step-management

154. A drug target is most closely related to
    1. the liver
    2. the cytochrome P 450 system
    3. a receptor site
    4. the circulatory system

155. How many "qhs" doses can be given "qd"?
    1. 24
    2. 4

3. 3
4. 1

156. What does the symbol s̄ mean?
    1. saturday
    2. standardize
    3. without
    4. skip

157. What is *MedWatch*?
    1. a service to monitor polypharmacy in the elderly
    2. a service to watch community pharmacies to prevent crime
    3. a reporting program available to health-care providers to report adverse events that can pose a serious health threat
    4. a device that signals the need for prescribed medication

I. Assisting the Pharmacist in Serving Patients

Dispensing Process

Prescriptions & Medication Orders

Labeling

Product Preparation

Specialty Activities

Information Resource

Safe Medication Practices

Common Ailments & Associated Drug Treatments

Pharmacokinetics

1. What steps should be taken when the prescriber's DEA number is missing from a prescription for a controlled substance?
    1. Leave it blank.
    2. Call the prescriber's office for verification and the DEA number.
    3. Enter a universal DEA number.
    4. Call the DEA for the prescriber's number.
2. What is meant by "ibuprofen 400 mg. PO QID c̄ food stat"?
    1. Give 400 mg. of ibuprofen with meals every day now.
    2. Give 400 mg. of ibuprofen with food 4 times a day and provide a status.
    3. Give ibuprofen 400 mg. with food every 4 hours.
    4. Give 400 mg. of ibuprofen orally with food 4 times a day starting immediately.

Dispensing Process

3. Use of abbreviations on a prescription label is limited to which of the following?
   1. Use abbreviations for commonly used words.
   2. Use abbreviations to conserve space on the label.
   3. Never use abbreviations.
   4. Use abbreviations to complete a heavy prescription workload timely.
4. Why is it important to have the full name of the patient on the prescription?
   1. The drug charge will go to the right party.
   2. The law requires the full name.
   3. This assures the right drug will be dispensed to the right patient.
   4. It is necessary for insurance purposes.
5. What is the meaning of "D/C ciprofloxacin start cephalexin 500 mg. PO QID × 14 D"?
   1. Stop ciprofloxacin and start cephalexin at 500 mg. by mouth 4 times a day for 14 doses.
   2. Stop ciprofloxacin and start cephalexin 500 mg. every 6 hours for 14 days.
   3. Discontinue ciprofloxacin. Start cephalexin 500 mg. orally 3 times a day for 14 doses.
   4. Discontinue the ciprofloxacin. Start cephalexin 500 mg. orally 4 times a day for 14 days.
6. What is the script?
   1. the prescription
   2. the counseling format
   3. the writing on the prescription blank
   4. the original hand-written order
7. What does a traditional prescription order provide?
   1. pricing
   2. the information needed to fill the medication order accurately
   3. third-party payor requirements
   4. a method to communicate the patient's illness
8. "MSO_4 0.002g IV push q2° prn for breakthrough pain for 9 h only" means what?
   1. Give 0.002g of magnesium sulfate through an intravenous push for breakthrough pain for only 9 hours.
   2. Give manganese sulfate 2 mg. pushing it in an intravenous fluid volume running every 2 hours for breakthrough pain for the next 9 hours.
   3. Give 2 mg. of morphine sulfate through an intravenous push every 2 hours when needed for breakthrough pain for up to 9 hours only.
   4. none of the above
9. The goal of distributive pharmacy can be summed up in which of the following ways?
   1. accurate inventory to assure ample drug stock
   2. counting, pouring, licking, and sticking

3. correct pricing for the medication
4. the right drug, for the right patient, at the right time

10. Which of the following auxiliary labels is appropriate for a prescription containing acetaminophen and codeine?
    1. MAY CAUSE PHOTOSENSITIVITY
    2. DRINK PLENTY OF FLUIDS
    3. MAY CAUSE DROWSINESS
    4. FINISH ALL THIS MEDICATION UNLESS OTHERWISE DIRECTED BY PRESCRIBER

11. What does "alprazolam 0.25 mg. PO x 1 stat" mean?
    1. Give alprazolam 25 mg. orally—provide one-time status.
    2. Give alprazolam 0.25 mg. tablet orally one-time immediately.
    3. Give alprazolam 0.25 mg. starting at one o'clock.
    4. Give alprazolam 0.25 mg. by mouth with first dose starting now.

12. Which strip label is very important to affix to all antibiotic prescriptions?
    1. MAY CAUSE DISCOLORATION OF URINE OR FECES
    2. TAKE WITH FOOD OR MILK
    3. FINISH ALL THIS MEDICATION UNLESS OTHERWISE DIRECTED BY PRESCRIBER
    4. THIS DRUG MAY IMPAIR THE ABILITY TO DRIVE OR OPERATE MACHINERY

13. What steps should you take when in doubt about the directions on a prescription?
    1. Use "as directed" for the directions.
    2. Tell the patient to check with the doctor.
    3. Call the doctor's office for verification of drug and clarification of directions.
    4. Do the best interpretation possible.

14. What does "D/C p.o. hydroxyzine. Give hydroxyzine 25 mg. IM q3-4h c̄ meperidine IV 75 mg. PRN pain" mean?
    1. Discontinue the oral hydroxyzine. Give hydroxyzine 25 mg. intramuscularly every 3-4 hours with meperidine 75 mg. intravenously as needed for pain.
    2. Stop the rectal hydroxyzine and give 25 mg. of hydroxyzine intramuscularly every 3 or 4 hours with meperidine 75 mg. when needed for pain.
    3. Stop hydroxyzine by mouth. Give 25 mg. of it intramuscularly over 3 to 4 hours with 75 mg. of meperidine as needed.
    4. Discontinue the oral hydroxyzine. Give 25 mg. of hydroxyzine in a muscle every 3 or 4 hours with meperidine #4 in 75 mg. doses as needed for pain.

15. What is the best way to provide additional special instructions to the patient that she or he will not forget?
    1. Affix a strip or auxiliary label on the container.
    2. Provide the patient with the reference page number.
    3. Tell the patient.
    4. All instructions must be typed on the label affixed to the container.

Dispensing Process

16. What information is not normally needed to fill a prescription?
    1. patient's name
    2. drug name and strength
    3. gender
    4. prescriber's name
17. How would you translate "Famotidine susp. 20 mg. pGT q. d."?
    1. famotidine sisps 20 mg. after gastric treatment every day
    2. famotidine suspension 20 mg. per drop every day
    3. famotidine suspension 20 mg. per gastric tube every day
    4. famotidine susceptibility 20 mg. per gastric test every day
18. You are unable to fill a prescription for a drug missing which of the following pieces of information?
    1. number of refills
    2. patient's age
    3. generic name of the drug
    4. dosage form
19. Which is the usual direction for administering p.o. tablets?
    1. insert
    2. apply
    3. take
    4. place
20. How would you explain the directions for Propoxyphene-N-100, 1 PO q4-6° prn HA?
    1. Take a tablet every 4 hours if needed for headache. Take 1 every 6 hours if not severe.
    2. Take one tablet orally every 4 to 6 hours when needed for headache.
    3. Take one tablet by mouth every 4-6 hours for hypertensive activity.
    4. Take 1 tablet every 4 or 6 hours when needed for headache.
21. Which of the following answers best describes what the prescription directions should contain?
    1. the name of the drug, how to take the drug, and the time
    2. the number of units, dosage form, frequency or specific time, and administration method
    3. the administration route, the number of units, the frequency, and the side effects
    4. the administration route, the frequency, the number of units, and the price
22. Which is the best way to describe in the directions the way to use an external ointment?
    1. apply
    2. insert
    3. instill
    4. take
23. Which is the best way to describe in the directions the way to take oral dosage forms?
    1. apply
    2. insert

3. instill
4. take

24. What are the directions for phenytoin 400 mg. p.o. stat, repeat in 4 hours, then 100 mg. PO q8h? (NOTE: Use phenytoin 100 mg. capsules)
    1. Take 4 capsules now and again in 4 hours. Then take 100 mg. every 8 hours.
    2. Take 4 capsules by mouth at once, then take 4 capsules 4 hours later, then take 1 capsule by mouth every 8 hours.
    3. Take 400 mg. now, 4 capsules in 4 hours, and 100 mg. orally every 8 hours.
    4. Take 8 capsules within 4 hours and follow with 1 capsule every 8 hours thereafter.

25. How many ccs are in a tuberculin syringe?
    1. 10 cc
    2. 5 cc
    3. 3 cc
    4. 1 cc

26. Which is the best way to describe in the directions the way to use eye drops?
    1. apply
    2. insert
    3. instill
    4. take

27. Which is the best way to describe in the directions the way to use rectal suppositories?
    1. apply
    2. insert
    3. instill
    4. take

28. Which of the following is the best direction to a patient for diphenhydramine 50 mg. p.o. q6h prn pruritis?
    1. Take one capsule by mouth every 6 hours when needed for itching.
    2. Take one pill every 6 hours as needed for sleeping problems.
    3. Take one every 6 hours for insomnia.
    4. Take one capsule orally every 6 hours when needed for seasonal disorder.

29. Which of the following routes of administration provides the fastest action?
    1. subq
    2. IM
    3. IV
    4. SL

30. A vaginal tablet is to a topical dosage form as a troche is to _____?
    1. oral dosage form
    2. topical dosage form
    3. inhalant dosage form
    4. parenteral dosage form

Dispensing Process

31. In addition to vials and ovals, other containers to package drugs for ambulatory patients include which of the following?
    1. ointment jars
    2. plastic bags
    3. dropper bottles
    4. answers 1 and 3

32. What are the directions for cefotetan 2g IV Q 12h x ii more doses?
    1. cefotetan 2 grams to IV grams every 12 hours for 2 more doses
    2. cefotetan 2 grams intravenously every 12 hours for 11 more doses
    3. cefotetan 2-IV grams every 12 hours for 11 more doses
    4. cefotetan 2 grams intravenously every 12 hours for 2 more doses

33. What does it mean to give a drug p. c.?
    1. through a personal catheter
    2. as a prepaid claim
    3. after meals
    4. before meals

34. If a prescriber wants only the brand name of the drug dispensed, how does the prescriber make this known?
    1. Check off or write "Dispense as written" or an equivalent designation on the prescription.
    2. Follow up the prescription with a personal call to the pharmacist.
    3. The prescriber's office follows up the prescription with a phone call to the pharmacy.
    4. The prescriber stamps the prescription with a red "B".

35. The label on a bulk bottle of the medication is important because it contains vital information such as what?
    1. the drug name
    2. the drug strength
    3. the expiration date of the drug
    4. all of the above

36. Which of the following items is not necessary on a prescription order?
    1. prescriber's name and title
    2. prescriber's age
    3. prescriber's office address
    4. prescriber's phone number

37. What should you do in the event of doubtful directions?
    1. Ask the patient.
    2. Leave it up to the pharmacist to check you.
    3. Clarify the directions before filling any prescription order.
    4. Make your best guess.

38. What are the directions for $MgSO_4$ 2 Gm. IV in 100 cc. NS over 1°?
    1. magnesium sulfate 2 grams intravenously in 100 cc. of normal saline over 1 hour
    2. morphine sulfate 2 grams intravenously in 100 cc. normal sulfate over 1 hour
    3. manganese sulfate 2 Gm. intravenously in 100 cc. of normal saline over 1 hour
    4. magnesium sulfate 2 milligrams in 100 cc. of Ringer's solution over 1 hour

39. In a weight-to-volume product of dextrose 50%, which of the following best describes the concentration of dextrose?
    1. The product is half dextrose.
    2. 50 mg. of dextrose in 100 ml. of water
    3. 50 Gm. of dextrose in 100 ml. of water
    4. 50 mg. of dextrose in 50 cc. of water

40. Never dispense ____.
    1. ointments with an "EXTERNAL USE ONLY" label
    2. a prescription with a red-bordered label
    3. capsules in a plastic vial
    4. guesswork

41. Should the patient's date of birth be a factor in drug selection by the doctor?
    1. yes
    2. maybe
    3. There is no relationship between age and drug selection.
    4. none of the above

42. In the hospital setting when reviewing a physician's order for drugs, in addition to the medications, strength, and frequency for use, what else is essential to review to assure the safety and well-being of the patient?
    1. non-pharmaceutical services performed
    2. legibility of the nurse's and doctor's signatures
    3. date of admission
    4. allergies or sensitivities to drugs and foods

43. What essential information should be on the label for a unit dose medication dispensed to a hospitalized patient?
    1. No information additional to the labeled information on the drug is necessary.
    2. patient's name, patient's room number and bed number, and date
    3. Only the patient's name is essential.
    4. Only the room and bed numbers are needed.

44. Why are strip labels necessary?
    1. They provide additional room for information that cannot be typed on the primary label.
    2. These labels are necessary to cover open area on the container.
    3. They provide additional information about using or taking the medication properly.
    4. Strip labels are required by law.

45. When do you need the pharmacist and pharmacy technician's initials on the prescription?
    1. The drug is filled by the technician and checked by the pharmacist.
    2. during work hours to acknowledge the presence of staff
    3. The pharmacist's and technician's initials are unnecessary.
    4. The initials are needed for controlled drug substances only.

Dispensing Process 125

46. What is the activity called whereby you enter the name of the drug on the prescription label?
    1. branding
    2. generic identification
    3. patent marking
    4. labeling

47. How would you begin the directions for a prescription for a rectal suppository?
    1. take
    2. instill
    3. apply
    4. insert

48. How would you start the directions for a prescription for a topical ointment or cream?
    1. take
    2. instill
    3. apply
    4. insert

49. What are the directions for meperidine in "Meperidine 50 mg. IM q 4h prn pain"?
    1. 50 mg. every 4 minutes as needed for pain
    2. 50 mg. intermittently in 4 hours when needed for pain
    3. 50 mg. intramuscularly every 4 hours as needed for pain
    4. 50 mg. injected every 4 hours as needed for pain

50. What is another way of writing twice daily after meals?
    1. BID pc
    2. B x D pc
    3. 2 xid pc
    4. BID ac

51. How would you start the directions for a prescription for nose drops?
    1. take
    2. instill
    3. apply
    4. insert

52. How would you start directions for a prescription for the oral route of administration?
    1. take
    2. instill
    3. apply
    4. insert

53. A prescription label must contain information that includes
    1. the prescriber's name, the prescriber's office staff who called in the refill, and the date of the next refill
    2. the patient's address, the patient's gender, the patient's age, and the patient's nationality

3. the payor's social security number, the insurance company name, the remaining deductible, and the co-insurance
4. the prescription number, the patient's full name, the directions for taking the medication, the prescriber's name, date, and name of the drug

54. Why is knowing the age of the patient important?
    1. The age may determine what dosage form should be used.
    2. The age may indicate if the dosage is appropriate.
    3. The age may pinpoint a condition contraindicated for the drug.
    4. answers 1 and 2

55. A prescription for a controlled substance must contain what required elements?
    1. the practitioner's office address
    2. the practitioner's DEA number
    3. the practitioner's signature
    4. all of the above

56. If no entry is made in the refills box, should you refill the drug?
    1. yes
    2. no
    3. maybe
    4. depends on the pharmacy policy

57. What are some indications that there has been tampering with the prescription?
    1. The written quantity of drug is smudged.
    2. The ink color or shade varies.
    3. The first name of the patient has been added where the prescriber wrote only the surname.
    4. answers 1 and 2

58. What are some important things to look for on the prescription blank?
    1. The name of the patient is clear and correct.
    2. The drug can be substituted therapeutically.
    3. The drug name and quantity are written clearly.
    4. answers 1 and 3

59. When is a prescriber's DEA number used?
    1. The DEA number is used if the drug is part of the Drug Evaluation Act.
    2. The DEA number is an arbitrary number that has little use in pharmacy practice.
    3. The DEA number is used to indicate the prescriber is a medical doctor and not an osteopath.
    4. The DEA number is required when the prescriber writes a prescription for a controlled substance.

60. The model prescription order contains which of the following elements?
    1. patient's name and address
    2. drug name, strength, and form
    3. quantity of the drug to be dispensed
    4. all of the above

Dispensing Process

61. What should be the primary concern of the pharmacy technician?
    1. The pharmacy is well-stocked at all times.
    2. The work area is clean.
    3. The safety and well-being of the patient is the primary concern.
    4. All records are maintained appropriately.
62. To what does the term "script" refer?
    1. the language used during counseling
    2. the profiled medical history for the patient
    3. the prescription
    4. the text used in journals
63. What types of practitioners can write prescriptions for legend drugs?
    1. physicians, osteopaths, dentists, veterinarians
    2. doctors of education, doctors of philosophy
    3. doctors of jurisprudence, psychologists, doctors of pharmacy
    4. none of the above
64. What is the purpose of having distilled water in the dispensing area?
    1. Distilled water is used to reconstitute medications requiring the addition of water.
    2. Distilled water is used for cleaning sinks and compounding countertops.
    3. Distilled water used to bathe patients with contagious diseases is supplied by the pharmacy.
    4. all of the above
65. Distributive pharmacy consists of a prescription order-processing activity and a medication product-preparation function. Which of the following functions does the pharmacy technician perform during the preparation phase?
    1. places orders for the required medications
    2. obtains patient health information
    3. retrieves the medication
    4. helps patrons with OTC needs
66. Which is not a drug delivery system?
    1. metered-dose inhalers (MDI)
    2. transdermal patch
    3. magnetic carrier transport
    4. ophthalmic drops
67. There may be evidence that any medication that interacts with erythromycin, ketaconazole, or itraconazole may also interact with which of the following fruit juices?
    1. orange
    2. apple
    3. grapefruit
    4. cranberry
68. Which term best describes an interaction in which one drug enhances the effect of another medication?
    1. antagonistic drug interaction
    2. synergistic drug interaction

3. pharmacokinetic drug interaction
4. none of the above

69. What technique is used to ease the withdrawal of a solution from a vial?
    1. Place an additional hole in the diaphragm next to the transfer site.
    2. Always make the withdrawal of solution in at least 2 steps.
    3. Inject a volume of air equal to the volume of solution for withdrawal into the vial before withdrawal.
    4. Shake vial immediately prior to withdrawing the solution.

70. Which term best describes an interaction in which one drug cancels out the effect of another medication?
    1. synergistic drug interaction
    2. additive drug interaction
    3. antagonistic drug interaction
    4. none of the above

71. Acetaminophen products containing codeine have the potential to cause what side effect?
    1. blurred vision
    2. drowsiness
    3. cough
    4. diarrhea

72. What is the "critical area" in an IV admixture program?
    1. area outside the hood
    2. the entire IV room
    3. the area near the door entering the IV room
    4. the area around the point of entry into the diaphragm of the vial

73. Since acetaminophen or a compound containing acetaminophen is metabolized in the liver, monitor closely if the patient is taking which of the following drug group(s)?
    1. alcohol
    2. tuberculosis drugs
    3. anticonvulsants
    4. all of the above

74. A primary reason why patients do not comply with the directions for taking medications is which of the following?
    1. The medication is not asthetically pleasing.
    2. The directions do not follow the patient's life style.
    3. The patient does not understand the directions.
    4. The patient takes the medication as he or she wants.

75. Hospital pharmacy practice deals primarily with what type of patient?
    1. ambulatory
    2. institutionalized
    3. adult day care
    4. clinic

76. What is characteristic of a solution?
    1. Solutions have specific colors.
    2. Solutes are suspended in a solvent.

3. All solutes are dissolved in the solvent.
4. Solutes in solution must have a specific surface area.

77. What is characteristic of a suspension?
    1. Solutes are dissolved in a thick solvent.
    2. All suspensions must be refrigerated.
    3. Suspension expiration dates are longer than those for solutions.
    4. Solutes are suspended in the solvent.

78. Community pharmacy practice deals primarily with what type of patient?
    1. institutionalized
    2. clinic
    3. ambulatory
    4. home infusion

79. What is the purpose of a buffer used in admixtures?
    1. It keeps ph from changing with the addition of acids or bases to the mixture.
    2. It prevents the drug from deteriorating the stomach lining.
    3. It extends the period of activity for the active ingredient.
    4. It assures sterility of the mixture.

80. Which of the following drugs works therapeutically by impacting on leukotrienes?
    1. zileuton
    2. zafirlukast
    3. zalcitabine
    4. answers 1 and 2

81. What is the most important thing to remember when preparing IV admixtures?
    1. Always use aseptic technique.
    2. Never forget to initial the label.
    3. Always run electrolytes separately into the evacuated container.
    4. Never use a plastic container when glass is available.

82. Why must you be especially careful of selling to a diabetic patient a drug in a syrup vehicle?
    1. Syrup preparations contain a high concentration of sugar.
    2. The sodium content enhances the effect of oral hypoglycemic agents.
    3. The alcohol content tends to decrease blood sugar.
    4. none of the above

83. Which of the following reasons best supports levigation in compounding a suspension?
    1. It decreases surface area.
    2. It enhances the viscosity of the preparation.
    3. It enhances the palatability of the preparation.
    4. It reduces particle size.

84. What is the major source of contamination in an IV admixture program?
    1. personnel
    2. materials and equipment

3. the environment
4. all of the above

85. Which of the following capsule sizes has the largest capacity?
    1. 000
    2. 1
    3. 4
    4. 5

86. Which of the following drug vehicles does not contain alcohol?
    1. elixir
    2. fluid extracts
    3. syrup
    4. tinctures

87. What is the purpose of using aseptic technique to prepare IV admixtures?
    1. to comply with JCAHO rules
    2. It is required by Boards of Pharmacy.
    3. It prevents contamination of the final product.
    4. to protect yourself from contact with toxic drugs

88. What method best accomplishes blending a powder into a cream base?
    1. sifting
    2. shaking
    3. levigation
    4. titrating

89. Which of the following is the most commonly used paper for weighing drugs on a balance?
    1. glassine
    2. parchment
    3. bond
    4. waxed

90. How long will 1L of an IV last that is running at 4.2 ml/minute?
    1. nearly 4 hours
    2. nearly 238 hours
    3. nearly 24 hours
    4. nearly 6 hours

91. Which of the following answers best describes the pharmacokinetics of a drug?
    1. ionization, mediation, and diffusion
    2. bioavailability, ionization, and transport
    3. absorption, distribution, and elimination
    4. polarity, molecular weight, and receptor affinity

92. Which are the primary organs involved in drug elimination?
    1. liver and kidney
    2. lungs
    3. spleen and gallbladder
    4. stomach and intestines

Dispensing Process

93. An order for an IV admixture includes the addition of 4,000 units of heparin. What volume from the 10,000u/ml vial will provide 4,000 u.?
    1. 0.5 ml
    2. 0.6 ml
    3. 0.04 ml
    4. 0.4 ml

94. What organ performs renal drug elimination?
    1. the skin
    2. the kidney
    3. the intestine
    4. the liver

95. What organ performs hepatic drug biotransformations?
    1. the kidney
    2. the spleen
    3. the liver
    4. the pancreas

96. What are the components of a traditional TPN?
    1. amino acids, lactated ringers, vitamins
    2. amino acids, dextrose, electrolytes, vitamins
    3. amino acids, normal saline, electrolytes
    4. amino acids, sterile water, lipids, electrolytes

97. What organ has responsibility for producing insulin?
    1. the pancreas
    2. the liver
    3. the kidney
    4. the adrenals

98. The term bioavailability best refers to which one of the following answers?
    1. the amount of drug detoxified in the liver
    2. the amount of drug remaining for therapeutic effect after elimination from the kidneys
    3. the rate and amount of therapeutically active drug that reaches systemic circulation
    4. the balance between the physical and chemical properties of a drug and the net therapeutic effect that remains

99. What is another name for the "fat" used in TPN or alone as an IV?
    1. triglycerides
    2. lipids
    3. cholesterols
    4. lipoproteins

100. Which route of drug administration provides the most rapid onset of drug effect?
    1. subcutaneous
    2. intramuscular
    3. buccal
    4. intravenous

101. The movement of drug molecules across a cell membrane from a region of high drug concentration to a region of low drug concentration is best known by which of the following answers?
    1. passive diffusion
    2. active transport
    3. bioavailability
    4. zero-order absorption
102. What should be done before clamping and cutting the transfer set attached to the finished IV product?
    1. Expel air from the flexible bag.
    2. Visually examine the final product for particles.
    3. Weigh the final product.
    4. Affix the label.
103. Biotransformation best refers to which of the following answers?
    1. drug absorption
    2. drug metabolism
    3. drug elimination
    4. drug distribution
104. What condition is most likely to use the HMG-CoA reductase inhibitor class of drugs?
    1. asthma
    2. hyperlipidemia
    3. hypertension
    4. diabetes
105. Which of the following is a common incompatability that results in a precipitant?
    1. potassium plus hydroxide salts
    2. magnesium plus sulfate salts
    3. calcium plus phosphate salts
    4. calcium plus acetate salts
106. What condition is most likely associated with leukotrienes?
    1. AIDS
    2. diabetes
    3. asthma
    4. depression
107. What component of a dosage form elicits the pharmacologic action?
    1. lubricant
    2. excipient
    3. binder
    4. active ingredient
108. What is TPN?
    1. Take pulse nightly.
    2. total parenteral nutrition
    3. test for potassium and nitrogen
    4. temperature and pulse normal
109. Why are there various types of dosage forms?
    1. to accomodate different ages and condition of patients
    2. to make the medication palatable

Dispensing Process

3. to permit maximum drug availability and absorption
4. all of the above

110. What is/are an example(s) of solid dosage forms?
    1. suspensions
    2. capsules
    3. lotions
    4. answers 1 and 3

111. What is needed to assure the prurity of a solution withdrawn from an ampule?
    1. visual inspection
    2. use a filter or filter needle for the transfer
    3. an in-date solution
    4. verification from the manufacturer

112. Typical topical dosage forms include which of the following?
    1. ointments
    2. creams
    3. transdermal patches
    4. all of the above

113. What could you expect a patient taking drugs for hypertension to be doing?
    1. maintaining a low-salt diet
    2. maintaining a low-carbohydrate diet
    3. maintaining a low-protein diet
    4. maintaining a "force-fluids" diet

114. Who traditionally decides which drugs will be part of the formulary?
    1. hospital upper management
    2. pharmacy and Nursing Workgroup
    3. the Pharmacy and Therapeutics Committee
    4. the Board of Pharmacy

115. Which triple-drug therapy is used to treat ulcers caused by Helicobacter pylori?
    1. omeprazole-amoxicillin-clarithromycin
    2. PEPTO-BISMOL—metronidazole—tetracycline or amoxicillin
    3. omeprazole-metronidazole-clarithromycin
    4. all of the above

116. Which directions should require clarification?
    1. Take medication three times a day.
    2. Ut. dict. or Take medication as directed.
    3. Take medication every day.
    4. Take medication at bedtime.

117. In general, what is a formulary?
    1. formulas used for compounding
    2. a carefully selected limited list of drugs
    3. drugs reimbursed by insurance companies
    4. a list of drugs used with disease management systems

118. What is a prevalent reason for medication errors?
    1. wrong patient
    2. wrong address

3. poor legibility
4. no refills listed

119. When is a pharmacy technician permitted to dispense guesswork?
    1. not to exceed 10% of the time
    2. The guesswork can be supported by logic.
    3. never
    4. within reason

120. Which of the following elements helps to determine what drugs are included in a formulary?
    1. the safety of the drug
    2. the margin of potential drug errors
    3. control of drug purchasing costs
    4. all of the above

121. Factors contributing to dispensing errors include
    1. excessive workload
    2. excessive interruptions
    3. inadequate training
    4. all of the above

122. Which error category is incorrect for categorizing errors?
    1. Category A—error with supporting reason
    2. Category B—error with no harm
    3. Category E—error with harm
    4. Category I—error resulting in death

123. Leukotrienes are most closely associated with which condition?
    1. leukemia
    2. asthma
    3. lipemia
    4. alcoholism

124. Another reference to arthritis may include which of the folowing?
    1. inflammation of the arterioles
    2. degenerative joint disease
    3. inflamed adrenal glands
    4. degenerative lung flexibility

125. Which of the following drugs would be expected as part of the pharmaceutical management for hemorrhoids?
    1. ibuprofen
    2. aminucuproic acid
    3. carisoprodol
    4. hydrocortisone

126. How would you characterize the organ responsible for circulatory congestion resulting from poor blood flow?
    1. peripheral digits
    2. eyes
    3. heart
    4. reproductive organs

127. Nonpharmaceutical management to treat chronic obstructive pulmonary disease should include which of the following?
    1. no smoking
    2. bland diet

3. heating pad
4. orthotics

128. How would you characterize the organ impacted by restricted pulmonary air flow?
    1. heart
    2. kidneys
    3. lungs
    4. liver

129. Nonpharmaceutical management to treat depression may include which of the following?
    1. UV therapy
    2. reclusion
    3. light therapy
    4. PT

130. Nonpharmaceutical management to treat fever should include which of the following?
    1. warm alcohol bath
    2. cool compresses
    3. high bulk diet
    4. pressed cucumber rub

131. Which of the following drugs would be expected as part of the pharmaceutical management for diarrhea?
    1. loperamide
    2. nizatidine
    3. phenytoin
    4. colchicine

132. Sumatriptan therapy is used for which of the following conditions?
    1. obesity
    2. nausea/vomiting
    3. migraine headaches
    4. vertigo

133. Which of the following drug types or drug classes is expected to appear in a pharmaceutical management regimen for constipation?
    1. laxatives
    2. antiemetics
    3. antinauseants
    4. expectorants

134. Nonpharmaceutical management to treat digestive tract ulcers should include which of the following?
    1. exercise program
    2. diet free from irritating foods
    3. psychotherapy
    4. low sodium diet

135. Which of the following drugs would be expected as part of the pharmaceutical management for epilepsy?
    1. lovastatin
    2. paclitaxol

3. allopurinol
4. hydantoin

136. Constipation primarily involves which area of human anatomy?
    1. cardiovascular system
    2. endocrine system
    3. respiratory system
    4. digestive system

137. Which of the following drugs would be expected as part of the pharmaceutical management for asthma?
    1. terazosin
    2. thioridazine
    3. terconazole
    4. theophylline

138. Arthritis primarily involves which area of the human anatomy?
    1. integumentary system
    2. skeletal system
    3. digestive system
    4. pulmonary system

139. Which of the following drug types or drug classes is expected to appear in a pharmaceutical management regimen for hemorrhoids?
    1. NSAIDs
    2. antiemetics
    3. stool softeners
    4. urinary acidifiers

140. Another reference to hypertension may include which of the following?
    1. anxiety
    2. high blood pressure
    3. hepatitis
    4. enteritis

141. Nonpharmaceutical management to treat congestive heart failure should include which of the following?
    1. physical therapy
    2. ice packs
    3. low sodium diet
    4. bland diet

142. Which of the following drug types or drug classes would be expected to appear in a pharmaceutical management regimen for congestive heart failure?
    1. endocrine drugs
    2. diuretics
    3. hormones
    4. antispasmodic drugs

143. Chronic obstructive pulmonary disease primarily involves which area of human anatomy?
    1. skeletal system
    2. endocrine system
    3. pulmonary system
    4. metabolic system

Dispensing Process

144. Which of the following drug types or drug classes is expected to appear in a pharmaceutical management regimen for chronic obstructive pulmonary disease?
    1. expectorants
    2. cardiovascular agents
    3. antidiarrheals
    4. antiemetic agents
145. Depression primarily involves which area of human anatomy?
    1. central nervous system
    2. respiratory system
    3. digestive system
    4. female reproductive system
146. How would you characterize the part of anatomy most contributing to an emotional condition of sadness and hopelessness?
    1. brain
    2. spinal column
    3. heart
    4. stomach
147. Which of the following drug types or drug classes is expected to appear in a pharmaceutical management regimen for fever?
    1. anti-inflammatories
    2. antineoplastics
    3. antipyretics
    4. anticoagulants
148. Nonpharmaceutical management to treat diarrhea whould include which of the following?
    1. ice packs
    2. stress management
    3. alcohol-free diet
    4. starchy foods
149. Which of the following drugs would be expected as part of the pharmaceutical management for constipation?
    1. disopyramide
    2. docusate
    3. dopamide
    4. disulfiram
150. What part of the anatomy is affected by the erosion of the lining of the digestive tract?
    1. esophagus
    2. stomach
    3. intestines
    4. all of the above
151. Which of the following drug types or drug classes can be expected to appear in a pharmaceutical management regimen for epilepsy?
    1. hypotensives
    2. hormones
    3. tranquilizers
    4. analgesics

152. Which of the following drug types or drug classes is expected to appear in a pharmaceutical management regimen for asthma?
    1. bronchodilators
    2. endocrine agents
    3. hormones
    4. antiemetic agents
153. Nonpharmaceutical management to treat arthritis should include which of the following?
    1. low-sodium diet
    2. specific exercise program
    3. smoking cessation
    4. humidifier
154. Of the following answers, which part of the anatomy is involved with hemorrhoids?
    1. stomach
    2. rectum
    3. duodenum
    4. jejunum
155. Hypertension primarily involves which area of the human anatomy?
    1. reproductive system
    2. endocrine system
    3. cardovascular system
    4. integumentary system
156. CHF primarily involves which area of the human anatomy?
    1. metabolic system
    2. endocrine system
    3. cardiovascular system
    4. integumentary system
157. Which of the following drugs would be expected as part of the pharmaceuctical management for congestive heart failure?
    1. indomethacin
    2. imipramine
    3. ibuprofen
    4. isosorbide
158. Another reference to chronic obstructive pulmonary disease may include which of the following?
    1. chronic bronchitis
    2. constipation
    3. incontinence
    4. thrombosis
159. Which of the following drugs would be expected as part of the pharmaceutical management for chronic obstructive pulmonary disease?
    1. isosorbide
    2. phenytoin
    3. albuterol
    4. colchicine

Dispensing Process

160. A patient may refer to depression as which of the following?
    1. guilt
    2. hostility
    3. blues
    4. anger

161. Which of the following drug types or drug classes is expected to appear in a pharmaceutical management regimen for depression?
    1. cardiovascular agents
    2. psychotherapeutic agents
    3. endocrine agents
    4. steroids

162. Which of the following drugs would be expected as part of the pharmaceutical management for fever?
    1. acetaminophen
    2. aspirin
    3. meperidine
    4. answers 1 and 2

163. Another reference by a patient to diarrhea may include which of the following?
    1. stooly
    2. flatulence
    3. the runs
    4. the burns

164. Which organ has a direct involvement on neurological seizure conditions?
    1. the eyes
    2. the brain
    3. the heart
    4. the intestines

165. Which of the following drug types or drug classes is expected to appear in a pharmaceutical management regimen for stomach ulcers?
    1. NSAIDs
    2. antacids
    3. ACE inhibitors
    4. CSF

166. Epilepsy primarily involves which area of human anatomy?
    1. central nervous system
    2. musculoskeletal system
    3. respiratory system
    4. digestive system

167. How would you characterize narrowing of the bronchioles in regard to its impact on an organ of the human anatomy?
    1. liver
    2. kidneys
    3. heart
    4. lungs

168. How would you characterize the part of anatomy affected by arthritic inflammation?
    1. joints
    2. skin
    3. muscles
    4. sinuses
169. Nonpharmaceutical management to treat hemorrhoids should include which of the following?
    1. increased dietary fiber
    2. warm compresses
    3. bland diet
    4. orthotics
170. Nonpharmaceutical management to treat hypertension should include which of the following?
    1. low-sodium diet
    2. low-fat diet
    3. exercise program
    4. all of the above
171. Which of the following drugs would be expected as part of the pharmaceutical management for lipid disorders?
    1. lovastatin
    2. pravastatin
    3. simvastatin
    4. all of the above
172. A patient may refer to myocardial infarctions by which of the following?
    1. heart attack
    2. infarct
    3. MI
    4. all of the above
173. Which of the following drugs would be expected as part of the pharmaceutical management for angina pectoris?
    1. nitroglycerin
    2. nitrofurantoin
    3. nizatidine
    4. nortriptyline
174. Another reference made to nausea and vomiting by patients may include which of the following?
    1. I'm throwing up.
    2. I'm sick to my stomach.
    3. I feel queezy.
    4. all of the above
175. Which of the following drugs could be expected to be part of the pharmaceutical management for atherosclerosis?
    1. loxapine
    2. lorezepam
    3. lovastatin
    4. loperamide

Dispensing Process

176. Which of the following drugs could be expected as part of the pharmaceutical management for depression?
    1. captopril
    2. diclofenac
    3. sertraline
    4. albuterol

177. Another reference to fever made by patients may include which of the following?
    1. temperature
    2. hallucinations
    3. constipation
    4. ringing in the ears

178. Which of the following drug types or drug classes is expected to appear in a pharmaceutical management regimen for diarrhea?
    1. vitamins
    2. electrolyte replacement fluids
    3. antineoplastics
    4. psychotherapeutic agents

179. Difficulty in passing stools impacts which organ of the human anatomy?
    1. kidney
    2. liver
    3. intestine
    4. stomach

180. Which of the following drugs could be expected as part of the pharmaceutical management for stomach ulcers?
    1. sucralfate
    2. spironolactone
    3. sulfisoxazole
    4. sulindac

181. Another common reference by patients to epilepsy may include which of the following?
    1. dropsies
    2. space outs
    3. foaming
    4. fits

182. Nonpharmaceutical management to treat asthma should include which of the following?
    1. heating pads
    2. massage
    3. remove irritants from living quarters
    4. high-fiber diet

183. Which of the following drug types or drug classes is expected to appear in a pharmaceutical management regimen for arthritis?
    1. TCA
    2. h_2 antagonist
    3. NSAIDs
    4. protease inhibitors

184. Hemorrhoids primarily involve which part of human anatomy?
    1. diverticulum
    2. hair follicles
    3. skin
    4. veins
185. How would you characterize abnormally high blood pressure in regard to its impact on an organ of the human anatomy?
    1. arteries
    2. blood
    3. intestines
    4. genitals
186. Which of the following drug types or drug classes is expected to appear in a pharmaceutical management regimen for hypercholesterolemia?
    1. antihypertensives
    2. antihyperlipidemics
    3. beta-blockers
    4. vitamin supplements
187. Myocardial infarctions primarily involve which area of human anatomy?
    1. metabolic system
    2. skeletal system
    3. endocrine system
    4. cardiovascular system
188. Which of the following drug types or drug classes is expected to appear in a pharmaceutical management regimen for angina pectoris?
    1. 5-HT$_3$ receptor agonists
    2. beta-blockers
    3. histamine H$_2$ blockers
    4. proton pump inhibitors
189. Nausea and vomiting primarily involve which areas of human anatomy?
    1. CNS and endocrine system
    2. CNS only
    3. CNS and respiratory system
    4. CNS and digestive system
190. Which of the following drug types or drug classes would you expect to appear in a pharmaceutical management regimen for atherosclerosis?
    1. HMG-CoA reductive inhibitors
    2. NSAIDs
    3. calcium channel blockers
    4. proton pump inhibitors
191. Which of the following is another way patients may refer to diabetes mellitus?
    1. sweet urine
    2. heavy urine
    3. sugar
    4. sweet blood

Dispensing Process

192. Which of the following drugs would be expected as part of the pharmaceutical management for pain?
    1. prochlorperazine
    2. propantheline
    3. propoxyphene napsylate/acetaminophen
    4. propranolol
193. Diarrhea primarily involves which area of human anatomy?
    1. nervous system
    2. endocrine system
    3. reproductive system
    4. digestive system
194. Asthma primarily involves which part of the human anatomy?
    1. metabolic system
    2. pulmonary system
    3. endocrine system
    4. digestive system
195. A patient may refer to gastroenteritis by which of the following?
    1. inflamed joints
    2. wheezing
    3. sweats
    4. stomach flu
196. Another reference to hemorrhoids may include which of the following?
    1. low blood
    2. piles
    3. fibroids
    4. stomach cramps
197. Which of the following drug types or drug classes is expected to appear in a pharmaceutical management regimen for hypertension?
    1. analgesics
    2. vitamin supplements
    3. vasodilators
    4. hormones
198. Nonpharmaceutical management to treat hypercholesterolemia should include which of the following?
    1. stress management
    2. restricted spice and acidic diet
    3. eliminate saturated fats from diet
    4. psychotherapy
199. Nonpharmaceutical management to treat myocardial infarctions should include which of the following?
    1. smoking cessation
    2. bland diet
    3. restricted exercise
    4. physical therapy
200. How would you characterize restricted coronary blood circulation in regard to its impact on an organ of the human anatomy?
    1. eyes
    2. reproductive system

3. heart and blood vessels
4. spleen

201. Which of the following drug types or drug classes is expected to appear in a pharmaceutical management regimen for nausea and vomiting?
    1. antiemetics
    2. laxatives
    3. vitamins
    4. psychotherapeutics

202. How would you explain to a customer asking about plaque formation in arteries which part of the body is involved?
    1. heart
    2. kidneys
    3. liver
    4. blood vessels

203. Diabetes mellitus primarily involves which area of human anatomy?
    1. respiratory system
    2. skeletal system
    3. reproductive system
    4. endocrine system

204. Which of the following drug types or drug classes is expected to appear in a pharmaceutical management regimen for pain?
    1. antiemetics
    2. antitussives
    3. analgesics
    4. anticoagulants

205. How would you characterize frequent bowel movements or loose stools in regard to its impact on a body part?
    1. the intestines
    2. the heart
    3. the lungs
    4. the kidneys

206. Gastroenteritis primarily involves which area of human anatomy?
    1. endocrine system
    2. integumentary system
    3. pulmonary system
    4. digestive system

207. Which of the following drugs would be expected as part of the pharmaceutical management for gastroenteritis?
    1. lomustine
    2. loperamide
    3. lorazepam
    4. loxapine

208. Which of the following drugs can be a part of the pharmaceutical management for hypertension?
    1. vecuronium
    2. vincristine

Dispensing Process

   3. verapamil
   4. valproic acid

209. Lipid disorders primarily involve which area of human anatomy?
   1. endocrine system
   2. pulmonary system
   3. cardiovascular system
   4. integumentary system

210. How would you characterize insufficient oxygenated blood in regard to its impact on an organ of the human anatomy?
   1. pancreas
   2. heart
   3. reproductive system
   4. kidneys

211. Nonpharmaceutical management to treat angina pectoris should include which of the following?
   1. aromatherapy
   2. environmental air conditioning
   3. orthotics
   4. low-fat diet

212. Which of the following drugs could be expected as part of the pharmaceutical management for nausea and vomiting?
   1. loperamide
   2. docusate
   3. prochlorperazine
   4. prazosin

213. Nonpharmaceutical management to treat atherosclerosis should include which of the following?
   1. low-fat diet
   2. weight-management program
   3. exercise program
   4. all of the above

214. Nonpharmaceutical management to treat diabetes mellitus should include which of the following?
   1. heating pads
   2. self-monitoring of BP
   3. restricted intake of simple sugars
   4. necessary orthotics

215. Which of the following drugs would be expected as part of the pharmaceutical management for an uncomplicated headache?
   1. amoxapine
   2. amiodarone
   3. acetaminophen
   4. atenolol

216. Another reference to constipation by patients may include which of the following?
   1. I'm gassy.
   2. I'm obstructed.
   3. I have rumbles.
   4. I'm nauseated.

217. Nonpharmaceutical management to treat gastroenteritis should include which of the following?
    1. hot/cold compresses
    2. fetal positioning
    3. electrolyte replacement
    4. raised legs
218. Which of the following drug types or drug classes would you expect to appear in a pharmaceutical management regimen for gastroenteritis?
    1. antiemetics
    2. anti-ulcer agents
    3. antineoplastic agents
    4. hypoglycemics
219. Which of the following drug types or drug classes may likely appear in a pharmaceutical management regimen for insomnia?
    1. steroids
    2. histamine H-2 antagonists
    3. antianxiety drugs
    4. antitussive drugs
220. Another reference made to lipid disorders by patients may include which of the following?
    1. runny sinuses
    2. cholesterol
    3. seizures
    4. migraines
221. Which of the following drug types or drug classes is likely to appear in a pharmaceutical management regimen for myocardial infarction?
    1. proton pump inhibitors
    2. colony stimulating factors
    3. antihyperlipidemics
    4. 5 HT-3 receptor agonists
222. Angina pectoris primarily inolves which area of human anatomy?
    1. cardiovascular system
    2. endocrine system
    3. metabolic system
    4. integumentary system
223. Another reference to atherosclerosis may include which of the following?
    1. multiple sclerosis
    2. hardening of the arteries
    3. muscular dystrophy
    4. arterial hypertrophy
224. How would you characterize, relative to a body part, the cause for insufficient metabolism of carbohydrates?
    1. the liver
    2. the pancreas
    3. the kidneys
    4. the lungs

225. Which of the following drugs could be expected as part of the pharmaceutical management for diabetes mellitus?
    1. glyburide
    2. fluoxetine
    3. amlodipine
    4. furosemide

226. Which of the following drug types or drug classes is likely to appear in a pharmaceutical management regimen for headaches?
    1. analgesics
    2. antiemetics
    3. ophthalmic drugs
    4. antitussives

227. Nonpharmaceutical management to treat constipation should include which of the following?
    1. heating pad
    2. ice packs
    3. increased dietary fibers
    4. massage

228. How would you characterize diverticulitis in regard to its impact on a body part?
    1. stomach
    2. rectum
    3. intestinal tract
    4. ureter

229. Which of the following drugs would you expect as part of the pharmaceutical management for insomnia?
    1. terfenadine
    2. pseudoephedrine
    3. belladonna
    4. zolpidem

230. Which of the following drugs is expected as part of the pharmaceutical management for myocardial infarction?
    1. nizatidine
    2. imipramine
    3. colchicine
    4. warfarin

231. What part of the human anatomy does atherosclerosis involve?
    1. cardiovascular system
    2. reproductive system
    3. digestive system
    4. endocrine system

232. Which of the following drug types or drug classes is expected to appear in a pharmaceutical management regimen for diabetes mellitus?
    1. ACE inhibitors
    2. hypoglycemic agents
    3. steroids
    4. immunosuppressive agents

233. Gastrointestinal ulcers primarily involve which area of human anatomy?
    1. integumentary system
    2. digestive system
    3. endocrine system
    4. nervous system

II Medication Distribution and Inventory Control Systems

Purchasing

Ordering

Vendors

Contracts

Inventory Control

Ordering Levels

Expiration Dates

Turnover

Preparation & distribution of inventory

Drugs

Invoice Processing

Pricing

1. What is the meaning of the last 2 digits in the NDC?
    1. They identify the manufacturer.
    2. They identify the brand/generic status.
    3. They identify the package size of the drug.
    4. They identify the list price.
2. What is the purpose of the sales invoice?
    1. It lists an order for specific merchandise from the supplier.
    2. It list slow-moving merchandise to be sold at sale prices.
    3. It lists the items shipped and the amount due to the supplier.
    4. It is an inventory sheet used to track merchandise.

3. What is a purchase order?
   1. a list of goods available for purchase
   2. an order form used to select items that a retailer wants to buy for resale
   3. an order form requesting the manufacturer's availability for selected merchandise
   4. the bill that accompanies the order
4. Who is the primary source of merchandise for pharmacy retailers?
   1. wholesalers and manufacturers
   2. store-to-store vendors
   3. larger discount stores
   4. warehouse clubs
5. Who pays the freight charges when shipping terms note "FOB Shipping Point"?
   1. seller
   2. transport company
   3. buyer
   4. post office
6. What special significance do the middle 4 numbers of the NDC have?
   1. They identify the manufacturer.
   2. They identify the product, form, and strength.
   3. They identify the product lot number.
   4. They identify the geographic region for distribution.
7. What is meant by the "terms of payment"?
   1. the date by which the buyer must pay for the merchandise
   2. how much the payment will be after deducting discounts
   3. the schedule for payment
   4. the remedies for late payments
8. What does the "extension" often found on the purchase order mean?
   1. a request for additional time to order merchandise
   2. an extended period to pay for merchandise
   3. Additional pages are needed to order quantities of merchandise.
   4. The quantity ordered times the unit price provides a total cost for the item ordered.
9. A purchase order usually contains the quantity, description, identifying number, and what else for each item ordered?
   1. gross price
   2. total charges
   3. unit price
   4. percent discount
10. Deducting the trade discount from the list price results in what type of price for the retailer?
    1. discount price
    2. net price
    3. retail price
    4. sales price

11. What is a trade discount?
    1. a reduction in the list price
    2. savings resulting from a trade between retailers
    3. a discount offered based on the type of trade
    4. a discount traded for services or merchandise
12. What special importance do the first 5 digits of the NDC have?
    1. They identify the manufacturer.
    2. They identify the product.
    3. They identify the lot number.
    4. They identify the year of manufacture.
13. What is another term for the suggested retail price?
    1. gross price
    2. trade price
    3. single price
    4. list price
14. Who pay the freight charges when shipping terms note "FOB destination"?
    1. buyer
    2. seller
    3. post office
    4. transport company
15. What is meant by "FOB" when identified in the shipping terms?
    1. free on board
    2. forward or backlog
    3. freeze original billing
    4. none of the above
16. How is the "net profit" determined?
    1. income less the deductions of all expenses
    2. reconciliation of income less checks written
    3. income less gross payroll amounts
    4. income less taxes paid
17. The NDC on bulk product labels is an acronym for what?
    1. new drug component
    2. national drug code
    3. new drug commodity
    4. national drug compendium
18. Which answer best describes how to purchase C-II drugs?
    1. C-II drugs may be ordered from the local wholesaler.
    2. Use a Schedule II DEA order form, file the order form and invoice separately from all other purchasing statements.
    3. C-II drugs may only be ordered from wholesalers that specialize in these drugs.
    4. Order C-II drugs according to the pharmacy's running inventory record for C-II drugs.
19. What does AWP represent when referring to drugs?
    1. the average wholesale price
    2. the actual wholesale price
    3. the average weighted price
    4. the actual weighted price

Preparation & Distribution of Inventory

20. What does the acronym DME mean?
    1. direct medication evaluation
    2. durable medical equipment
    3. drug measuring equipment
    4. drug measure estimate

21. What is a physical inventory?
    1. the health status of employees
    2. the physical condition of the pharmacy location
    3. accountability of the actual items in stock
    4. the Board of Pharmacy requirement for compounding equipment

22. What is a periodic inventory?
    1. a monthly accountability of sales activity
    2. accountability for merchandise taken monthly, quarterly, semi-annually, or annually
    3. taxes paid semiannually
    4. accountant's reference to a balance sheet

23. What is a perpetual inventory system?
    1. a constant record of items purchased and sold
    2. a record of items for purchase
    3. a record of slow-moving items
    4. sales merchandise

24. What is inventory?
    1. stock that is sale-priced
    2. excess merchandise
    3. merchandise offered for resale
    4. merchandise required to be ordered

25. What is inventory turnover?
    1. the number of times specific items are replaced during a given period
    2. replacement of old items with different items
    3. the number of times a sale must be held to reduce slow-moving stock
    4. the number of new drugs that replace or add to similar types of drugs

26. What system is used to identify all prescription drugs approved by the FDA?
    1. UPS
    2. UPN
    3. NDC
    4. UPC

27. How can most products be identified for inventory?
    1. SKI System
    2. UPC System
    3. UPN System
    4. UPS System

28. Which of the following answers best defines a stock bottle?
    1. a bulk product manufactured in house
    2. a bulk bottle

3. fast moving product
   4. product support by backup stock
29. What is the term that notes the amount a business pays for merchandise, including freight changes and services?
   1. selling price
   2. markup
   3. cost
   4. profit
30. What does the term markup mean?
   1. the price charged to best customers
   2. contract prices
   3. the amount added to the cost of a product
   4. entries to sales records
31. What is the markup formula?
   1. selling price + cost = markup
   2. markup − cost = selling price
   3. cost − markup − discount = selling price
   4. cost + markup = selling price
32. What is meant by retail?
   1. opening a pharmacy
   2. selling goods directly to the customer
   3. reselling goods to institutions
   4. the profit made above wholesale
33. What is wholesale?
   1. engaging in the whole resale of goods to community pharmacies, manufacturers, and the public
   2. the sale of goods at a lower price to retailers who resell the goods to customers at a higher price
   3. the selling of goods in "whole" lots only
   4. the price the manufacturer charges the public if sold directly
34. What is it called when the expenses exceed the revenues?
   1. break-even
   2. income
   3. loss
   4. none of the above
35. What is COGS?
   1. counting overstocked goods for sale
   2. cleaning of goods on shelves
   3. cost of goods sold
   4. none of the above
36. What is a markdown?
   1. revenue excluded from the purchase price
   2. a percentage reduction in price
   3. a professional courtesy extended to other pharmacy practitioners
   4. a pricing practice used to keep prices stable
37. What are rebates?
   1. a return of money for meeting specific conditions
   2. a finance charge for late payments

Preparation & Distribution of Inventory 153

       3. specific dates by which payments must be made
       4. additional goods supplied as an incentive to purchase larger quantities

38. Which of the following is an important element to look at on the bulk drug label of an incoming order?
    1. the condition of the label
    2. the name of the drug manufacturer
    3. the expiration date of the drug
    4. none of the above
39. Why is it important to rotate stock on shelves?
    1. prevents dust build-up
    2. saves shelf space
    3. assures use of the product with the closest expiration date.
    4. provides a busy image
40. What is the difference between pricing and reimbursement?
    1. Pricing drugs is determined by the retailer, and reimbursement is determined by the third-party payer.
    2. Pricing refers to all items and services, while reimbursement refers only to drugs.
    3. Pricing and reimbursement are the same
    4. third-party payers price, and retailers receive reimbursement.

III Operations

Management

    Policies

    Retailing

Facilities & Equipment

    Maintenance

    Records

Human Resources

    Roles, Duties, and Responsibilities

Information

    Electronic

    Computer

Hardcopy

Sources

Law

Statutes

Regulations

Standards

Ethics

Communications

1. A pharmacy department should be able to provide which of the following management essentials to guide pharmacy activities?
   1. a drug price list
   2. a mission statement
   3. a formulary
   4. a list of pharmaceutical representatives
2. What is the allowable charge?
   1. The allowable charge is the maximum amount that a third-party payer will reimburse a provider or supplier for a service or an item.
   2. The allowable charge is the amount that is charged to the patient.
   3. The allowable charge is the amount applied to a patient's credit card.
   4. The allowable charge is the least amount that a third-party payer will reimburse for an item or service.
3. Basic records for pharmacy include which of the following?
   1. purchase orders
   2. shipping invoices
   3. narcotic forms
   4. all of the above
4. Which of the following is the best description of a salary?
   1. an hourly pay for a specific set of hours worked
   2. variable payments made to employees working split shifts
   3. a contract to keep a payment to an employee set for a specified period of time
   4. a preset amount paid to an employee
5. What should be the foremost purpose of management?
   1. to provide guidance to staff while also motivating them
   2. to implement a Total Quality Environment
   3. to monitor Continuous Quality Improvement
   4. to establish teams and workgroups

Communications

6. What is the best description of an employee's wage?
   1. a contract payment for the work performed by an individual over a period of time
   2. an accounting term used to identify the difference between expenses and income
   3. payment made to an employee for services performed
   4. the amount subtracted for taxes and other deductions

7. What is an employee's take-home net pay?
   1. the employee's pay reported to the Internal Revenue Service
   2. the pay after income taxes and other deductions are subtracted from gross pay
   3. the portion of an employee's pay set aside for unseen circumstances
   4. new employee tax

8. What does it mean to extend a professional courtesy?
   1. saying "hello" to other health-care staff
   2. providing a discount for items purchased by a person in the same field of work
   3. a daily greeting for patients
   4. none of the above

9. What is a payroll?
   1. a list showing names of employees and the amounts paid to each for work performed
   2. the total wages paid out to working staff
   3. a list of full- and part-time employees
   4. a form listing employees submitted to the Internal Revenue Service

10. Who is the person in whose name an insurance policy is held?
    1. dependent
    2. subscriber
    3. beneficiary
    4. patient

11. What method is used to assign budget money for various operations in the pharmacy?
    1. inventory
    2. budgetary balancing
    3. departmental accountability
    4. allocation

12. What does "overhead" include?
    1. lighting
    2. rent, insurance, and utilities
    3. ceilings
    4. the cost for contracts with manufacturers

13. What is the importance of having a budget?
    1. It is required by the Internal Revenue Service.
    2. It plans the pharmacy's future income, costs, expenses, and net income.
    3. It is useful in purchasing merchandise on credit.
    4. It indicates the status of a company's ability to operate.

14. How does the pharmacy receive payment from third-party payers?
    1. The pharmacy submits a purchase order to the third-party payer.
    2. The pharmacy submits a Certificate of Medical Necessity to the third-party payer.
    3. The pharmacy submits a claim form to the third-party payer.
    4. The pharmacy telephones information to the third-party payer.
15. What does a financial statement provide?
    1. the solidness of a company's sales
    2. the assets a company has in savings
    3. a status of a company's progress and financial condition
    4. a verification of assets required by the IRS
16. What tools does management use to analyze business performance?
    1. income statements and balance sheets
    2. owner's equity and liability
    3. purchasing invoices
    4. inventory on hand
17. Unless otherwise noted, how should drugs be stored?
    1. in a non-humid area at room temperature
    2. on shelves
    3. in a medicine cabinet
    4. any place convenient to enable efficient filling and dispensing of drugs
18. The current pharmacy budget is $118,000. What is the new budget resulting from an 11-percent increase?
    1. $130,780
    2. $130,980
    3. $12,980
    4. $125,000
19. What is the term that expresses the revenue in excess of the cost of the merchandise plus the operating expenses?
    1. markup
    2. profit
    3. net revenue
    4. earnings
20. Some examples of third-party payers include which of the following?
    1. medicare
    2. medicaid
    3. insurance companies
    4. all of the above
21. The pharmacy budget for drug purchasing is $85,000. The total pharmacy budget is $112,000. What percent of the budget is devoted to drug purchases?
    1. 24%
    2. 76%
    3. 32%
    4. 85%

Communications

22. An important part of a pharmacy operations management includes which of the following?
    1. workflow
    2. monitoring personal leave
    3. compliance with overall facility needs
    4. providing whatever upper management wants

23. Which of the following ways best keeps pharmacy objectives uniform and consistent?
    1. a newsletter
    2. e-mail
    3. journals
    4. a current policy manual

24. What are the two methods of pharmacy claims submission to third-party payers?
    1. U.S. Postal Service and Fed Ex
    2. hardcopy and electronic
    3. UPS and DHL
    4. fax and e-mail

25. How does the pharmacy computer connect to other computers by means of the telephone?
    1. modem
    2. initialization
    3. tutorial programs
    4. CPU

26. What does the gauge of a needle denote?
    1. weight
    2. length
    3. outside diameter
    4. shaft thickness

27. Computers work to do which of the following tasks?
    1. input and output
    2. processing and storage
    3. communications
    4. all of the above

28. What is the benefit of having a facility accredited by an accrediting agency?
    1. An accredited facility has a line of credit.
    2. Accreditation provides official approval to a facility that it meets a set of specific standards.
    3. Accreditation entices more patients to come to the facility.
    4. There is no benefit to being accredited.

29. What is the *terminal* when referring to computers?
    1. a stand-alone computer
    2. the last entry made to a computer
    3. workstation where data is entered or received
    4. specialized terms used to command computer functions

30. Which sanitizer is commonly used to clean the work surface of the laminar flow hood?
    1. 70% isopropyl alcohol
    2. povidone–iodine solution
    3. 95% ethyl alcohol
    4. hexachlorophene

31. What directs the computer to do specifically designed functions?
    1. the menu
    2. the disk drive
    3. function keys
    4. software

32. What are the special words and symbols called that will perform tasks on the computer?
    1. menu
    2. commands
    3. syntax
    4. graphic utilization integrators

33. What are the parts of a syringe?
    1. reservoir and plunger
    2. cylinder and piston
    3. barrel and plunger
    4. container and head

34. What would be the need for a *fast mover* or *speed shelf*?
    1. Reordering drugs is faster.
    2. These shelves are easily removed and placed somewhere else.
    3. Frequently used medications are conveniently accessible.
    4. This area contains only controlled substances for better monitoring.

35. Which of the following is important to have included in a chemo prep kit?
    1. dust/mist respirator
    2. gown
    3. gloves
    4. all of the above

36. As a group, what is a keyboard, monitor, printer, and CPU called?
    1. input devices
    2. computer hardware
    3. output devices
    4. peripherals

37. What piece of equipment reduces the risk of contamination when preparing IV admixtures?
    1. UV lighting
    2. humidifiers
    3. laminar flow hoods
    4. foot-pedal sinks

38. What is a special secured word to activate a computer called?
    1. tutorial
    2. password

3. sign on
4. alpha-numeric byte

39. Why have a tachy mat at the entrance of an IV room?
    1. to post notes for other staff
    2. to reduce particle counts from shoe debris
    3. to leave shoes on while in the IV room
    4. to soften the hard entry to the IV room

40. Which of the following uses is not usual for a spatula?
    1. removing cotton from a stock bottle of medication
    2. mixing ointments
    3. counting tablets and capsules
    4. transferring powders from the stock bottle to the balance

41. In a typical computer setup, which piece of hardware is the output device?
    1. printer
    2. monitor
    3. modem
    4. keyboard

42. What part of the laminar flow hood is responsible for removing microcontaminents?
    1. the prefilter
    2. the plenum
    3. the recovery vent
    4. the HEPA filter

43. Computer technology in pharmacy has had its greatest impact on which of the following?
    1. communications
    2. record keeping
    3. distributive pharmacy processing efficiency
    4. all of the above

44. What is essential in order to be prepared for handling cytotoxic drug spills?
    1. latex gloves
    2. disposable, absorbent towels
    3. chemo spill kits
    4. alternate pharmacy workplace

45. What basic areas do typical pharmacy settings have?
    1. a work area, a bathroom, and a coatroom
    2. a work area and a storage area
    3. a storage area and a counseling area
    4. a drug area and a durable medical equipment (DME) area

46. What equipment is available to measure out liquid volumes?
    1. counting trays
    2. graduates
    3. ovals
    4. vials

47. What is a hard copy?
    1. the file on the hard drive
    2. an acetate copy used for overhead projectors

3. rigid backing for a presentation
4. information printed on paper

48. On a computer, what are the keys used to perform selective functions?
   1. macro keys
   2. command keys
   3. function keys
   4. icons

49. What type of equipment would you expect to find in the work area?
   1. medications, containers, and insurance forms
   2. a balance, a laminar flow hood, and backup stock
   3. only containers and IV bottles
   4. graduates, beakers, and counting trays

50. What would you likely do to prevent losing information stored on a computer?
   1. Make a backup disk.
   2. File the hard copy of the information.
   3. Save a record file to the hard drive.
   4. Never shut off the computer.

51. What part of the computer is the CPU?
   1. cleared processing unit
   2. central processing unit
   3. continuous production unit
   4. central production unit

52. What is one of the most important elements in measuring a technician's performance?
   1. size of the pharmacy
   2. competencies in pharmacy tasks
   3. work load
   4. size of formulary

53. Important professional attributes include which of the following?
   1. religious beliefs
   2. economic standing
   3. knowledge of the scope and limitations of pharmacy practice
   4. political involvement

54. Which duties are appropriate to include in a position description for a pharmacy technician?
   1. receiving the prescription, interpreting questionable items on the prescription, selecting the drug, labeling the container, and giving the prescription to the patient.
   2. receiving the prescription, assessing the contents of the prescription, selecting the appropriate drug, labeling the container, and counseling the patient
   3. receiving the prescription, reading the contents of the prescription, selecting the appropriate drug, labeling the container, and notifying the pharmacist that the drug is ready to be checked
   4. receiving the prescription, reading the contents of the prescription, substituting the drug as needed, labeling the container, and notifying the pharmacist that the drug is ready for approval

Communications

55. What are some of the specific criteria that define the pharmacy technician's role?
    1. works under the supervision of a pharmacist
    2. does not require the use of professional judgement
    3. performs duties primarily limited to the distributive pharmacy function
    4. all of the above

56. To whom does the term pharmacy "supportive personnel" refer?
    1. all personnel who support the pharmacy activity
    2. only personnel in the pharmacy department who support the pharmacy activity
    3. only pharmacy technicians
    4. there is no such reference to personnel

57. A pharmacy technician may perform the following duties except
    1. maintaining patient records, filling and dispensing routine orders for stock supplies, and preparing routine compounding
    2. pre-packaging drugs in single dose or unit-of-use, maintaining inventories of drug supplies, and completing insurance forms
    3. providing clinical counseling, providing clinical information to medical staff, and providing patient medical history data to requestors
    4. labelling medication doses, preparing intravenous admixtures, and maintaining the cleanliness of laminar flow hood

58. Pharmacy technician duties vary with the type of pharmacy practice. However, the duties common to all pharmacy settings include
    1. preparing ophthalmic preparations under the laminar flow hood
    2. measuring out the medication, labeling the container, and packaging the medication
    3. helping patients find over-the-counter products
    4. advising patients about symptoms they are experiencing

59. "Studies in animals or human beings have demonstrated fetal abnormalities or there is evidence of fetal risk based on human experience, or both, and the risk of the use of the drug in pregnant women clearly outweighs any possible benefit. The drug is contraindicated in women who are or may become pregnant" defines which category of fetal risk factors?
    1. X
    2. A
    3. B
    4. C

60. "There is positive evidence of human fetal risk, but the benefits from use in pregnant women may be acceptable despite the risk (e.g., if the drug is needed in a life-threatening situation or for a serious disease for which safer drugs cannot be used or are ineffective)" defines which category of fetal risk factors?
    1. X
    2. B
    3. C
    4. D

61. "Either studies in animals have revealed adverse effects on the fetus (teratogenic or embryocidal, or other) and there are no controlled studies in women, or studies in women and animals are not available. Drugs should be given only if the potential benefit justifies the potential risk to the fetus" defines which category of fetal risk factors?
    1. A
    2. X
    3. C
    4. D
62. "Either animal-reproduction studies have not demonstrated a fetal risk but there are no controlled studies in pregnant women, or animal-reproduction studies have shown an adverse effect (other than a decrease in fertility) that was not confirmed in controlled studies in women in the first trimester (and there is no evidence of a risk in later trimesters)" defines which category of fetal risk factors?
    1. A
    2. B
    3. C
    4. X
63. "Controlled studies in women fail to demonstrate a risk to the fetus in the first trimester (and there is no evidence of a risk in later trimesters) and the possibility of fetal harm appears remote" defines which category of fetal risk factors?
    1. A
    2. B
    3. C
    4. D
64. Which of the first letters of the FDA's basic rating system indicates a drug meets bioequivalence requirements?
    1. A
    2. B
    3. E
    4. none of the above
65. What does the *Orange Book* code designated as "AA" mean?
    1. no bioequivalence problems in conventional dosage forms
    2. documented bioequivalence problem
    3. meets necessary bioequivalence requirements
    4. testing standards are insufficient for determination
66. What does the *Orange Book* code designation "BD" mean?
    1. no bioequivalence problems in conventional dosage forms
    2. documented bioequivalence problem
    3. meets necessary bioequivalence requirements
    4. potential bioequivalence problem
67. What does the *Orange Book* code designation "AB" mean?
    1. no bioequivalence problems in conventional dosage forms
    2. documented bioequivalence problem
    3. meets necessary bioequivalence requirements
    4. potential bioequivalence problem

Communications

68. What does the *Orange Book* code designation "BP" mean?
    1. no bioequivalence problems in conventional dosage forms
    2. documented bioequivalence problem
    3. meets necessary bioequivalence requirements
    4. potential bioequivalence problem
69. What does the *Orange Book* code designation "BS" mean?
    1. no bioequivalence problems in conventional dosage forms
    2. documented bioequivalence problem
    3. meets necessary bioequivalence requirements
    4. testing standards are insufficient for determination
70. What does the *Orange Book* code designation "B*" mean?
    1. no bioequivalence problems in conventional dosage forms
    2. documented bioequivalence problem
    3. meets necessary bioequivalence requirements
    4. requires further FDA investigation and review
71. What does the *Orange Book* code designation "BX" mean?
    1. no bioequivalence problems in conventional dosage forms
    2. insufficient data to confirm bioequivalence
    3. meets necessary bioequivalence requirements
    4. potential bioequivalence problem
72. What does the acronym ASHP represent?
    1. american Society of Hospital Pharmacists
    2. american Schools of Health Practices
    3. association of Specialty Health Practitioners
    4. american Society of Health-Systems Pharmacists
73. Which association best represents the interests of chain drug stores?
    1. NABP
    2. APHA
    3. NCPA
    4. NACDS
74. To what does FDA refer?
    1. Food and Drug Administration
    2. Federal Drug Advocates
    3. fosinopril-digoxin-atenolol
    4. food drugs and accessories
75. Which of the following information sources provides the best overall drug information?
    1. *The Merck Manual*
    2. *Dorland's Medical Dictionary*
    3. *The PDR*
    4. answers 1 and 2
76. A Class IV controlled substance is represented by which of the following?
    1. diazepam
    2. propoxyphene
    3. zolpidem
    4. all of the above

77. Which of the following practitioners is not authorized to write for legend drugs?
    1. LSW
    2. MD
    3. DDS
    4. DO
78. A health-care profession's code of ethics should stress which of the following?
    1. a patient's dignity
    2. confidentiality
    3. appropriate payment for services
    4. answers 1 and 2
79. What is a promissory agreement between two or more persons or entities?
    1. contract
    2. consent
    3. precedent
    4. standard of care
80. How many refills are permitted for Class III controlled substances?
    1. 0
    2. 5 renewals within 6 months
    3. 6 renewals within 1 year
    4. 1 per month with prescriber's verification
81. What is the single term for legislative enactments that declare, command, or prohibit something?
    1. suits
    2. subpoenas
    3. torts
    4. statutes
82. Which of the following answers best describes statements made to a person in a position of trust, such as an attorney or health-care practitioner?
    1. confidentiality
    2. privacy
    3. privileged communication
    4. reasonable care
83. What definition best describes malpractice?
    1. the legal concept of cause and effect
    2. professional negligence, improper discharge of professional duties, or a failure of a professional to meet the profession's standard of care which results in harm to another
    3. acts performed or omitted that a reasonably prudent health-care practitioner would have done or not done under the same or similar situation
    4. the legal rights and duties of private persons
84. In which controlled substances class will you find meperidine?
    1. Class II
    2. Class III

Communications

3. Class IV
4. Class V

85. What is a suit?
    1. a court order that requires a person to appear in court to give testimony
    2. a court proceeding that seeks damages and other legal remedies
    3. a legal or civil wrongdoing committed by a health-care practitioner
    4. a previous decision that serves as the authority for a similar case

86. What is a subpoena?
    1. a court order that requires a person to appear in court to give testimony
    2. a court proceeding that seeks damages and other legal remedies
    3. a legal or civil wrongdoing committed by a health-care practitioner
    4. a previous decision that serves as the authority for a similar case

87. What does confidentiality mean to the pharmacy technician?
    1. maintaining the privacy of a patient's medical history
    2. keeping a pharmacy technician's salary private
    3. the pharmacy supervisor having confidence in the pharmacy technician
    4. being able to complete third-party payer forms in any environment

88. What is the function of the State Board of Pharmacy?
    1. to regulate pharmacy practice
    2. to inspect pharmacies for compliance with pharmacy laws and regulations
    3. to revoke or suspend the pharmacy license of practitioners found guilty of violating pharmacy laws or regulations
    4. all of the above

89. In which controlled substances class will you find morphine?
    1. Class V
    2. Class IV
    3. Class III
    4. Class II

90. Which of the following answers best describes how scheduled prescriptions are maintained?
    1. separate file for each schedule
    2. separate files for Schedule II and Schedules III through V
    3. schedule III through V prescriptions filed with all other prescriptions, but identified with a red "C" stamped in the lower right-hand corner of the prescription
    4. any of the above filing systems

91. What does the legend on the label of a stock bottle contain?
    1. "Warning: May be habit forming."
    2. "Caution: Federal law prohibits dispensing without a prescription."
    3. "Note: Original date of manufacture is *MO/DAY/YR*."
    4. "Certification: Lot number and expiration date have been verified and validated."

92. When can the generic form of a drug be produced?
    1. anytime
    2. when the brand-name patent expires
    3. FDA determines production for generics
    4. generic drug manufacturers determine when to produce generic drugs

93. In which class of controlled substances will you find oxycodone?
    1. Class I
    2. Class II
    3. Class III
    4. Class IV

94. Which answer best describes how often a C-III and C-IV drug may be refilled?
    1. PRN
    2. All refills must be called in to the prescriber and verified.
    3. one refill every 30 days for up to 1 year
    4. Five refills within six months may be authorized by the prescriber.

95. Which answer best describes the information that should be included in a Schedule V sales log for C-V drugs sold without a prescription?
    1. dispensing date, printed name, signature and address of buyer, name and quantity of the product sold, and the pharmacist's signature
    2. dispensing date, signature and phone number of buyer, the product name, and the product company with lot number
    3. dispensing date, signature of buyer, name and quantity of product sold, price of the product sold, and the lot number of the product sold
    4. dispensing date, buyer signature, pharmacist signature, product name and amount, and the expiration date of the product

96. Which of the following answers best describes how a Schedule II drug is dispensed?
    1. verbal or written prescription
    2. only on written prescriptions signed by the prescriber
    3. a verbal order is appropriate to comply with the law
    4. verbal, written, or faxed prescription complies with the law

97. What does Schedule I tell us about a drug?
    1. The drug has a high potential for abuse and physical or psychological dependence.
    2. The drug has a moderate or low potential for physical dependence and a high potential for psychological abuse.
    3. The drug has limited potential for physical or psychological dependence.
    4. The drug has no currently accepted medical use in the United States because of its high potential for abuse.

98. How many renewals are permitted for a Class II controlled substance?
    1. 0
    2. 1
    3. 2
    4. 3
99. Which answer best describes when a Schedule II prescription may be honored as a verbal order?
    1. never
    2. in an emergency
    3. There are no stipulations as long as you can acknowledge the prescriber by his/her DEA number.
    4. in an emergency as long as only the required quantity is dispensed and the prescriber follows up with a written and signed prescription within 72 hours
100. How long are you permitted to complete a partial filling of a Schedule II prescription?
    1. 24 hours
    2. 48 hours
    3. 72 hours
    4. A new prescription is needed for the remainder at any time.
101. What does C-III tell us about a drug?
    1. The drug has a high potential for abuse and physical or psychological dependence.
    2. The drug has a moderate or low potential for physical dependence and a high potential for psychological abuse.
    3. The drug has limited potential for physical or psychological dependence.
    4. The drug has a limited potential for abuse.
102. What does C-II tell us about a drug?
    1. The drug has a high potential for abuse and physical or psychological dependence.
    2. The drug has a moderate or low potential for physical dependence and a high potential for psychological abuse.
    3. The drug has limited potential for physical or psychological dependence.
    4. The drug has a limited potential for abuse.
103. What are the renewal requirements for Class IV prescriptions?
    1. not renewable
    2. renew as needed
    3. 5 renewals within 6 months
    4. 6 renewals within 1 year
104. What does C-IV tell us about a drug?
    1. The drug has a high potential for abuse and physical or psychological dependence.
    2. The drug has a moderate or low potential for physical dependence and a high potential for psychological abuse.

3. The drug has limited potential for physical or psychological dependence.
4. The drug has a limited potential for abuse.

105. What does C-V tell us about a drug?
    1. The drug has a high potential for abuse and physical or psychological dependence.
    2. The drug has a moderate or low potential for physical dependence and a high potential for psychological abuse.
    3. The drug has limited potential for physical or psychological dependence.
    4. The drug has a limited potential for abuse.

106. How are controlled substances classified by the federal law?
    1. "N" for narcotics and "CS" for controlled substances
    2. schedules I through V
    3. registry I through V
    4. Controlled Substances Index (CSI)

107. Which agency regulates controlled substances?
    1. The Food and Drug Administration
    2. The Public Health Service
    3. The Drug Enforcement Agency
    4. The Bureau of Narcotics and Dangerous Substances

108. How long is a prescriber given to confirm in writing verbal prescriptions for Class II controlled substances?
    1. 24 hours
    2. 2 days
    3. 72 hours
    4. 1 week

109. ERYTHROCIN is to a brand name as erythromycin is to what?
    1. proprietary name
    2. patent name
    3. generic name
    4. trademark name

110. Which of the following products is/are exempted from being dispensed in child-resistant containers?
    1. diltiazem
    2. nitroglycerin sublingual tablets
    3. atenolol
    4. answers 1 and 3

111. What should you do if you provide a non-controlled substance copy of a prescription to another pharmacy?
    1. it depends on the State's pharmacy regulations.
    2. Cancel the original prescription.
    3. Write up a new prescription.
    4. Enter the date and pharmacy you are providing the prescription and your initials.

112. Which of the following is an example of a Class II controlled substance?
    1. chloral hydrate
    2. paregoric

Communications

    3. fentanyl
    4. diazepam

113. Which of verbal, faxed, and in-person prescriptions can be handled by the pharmacy technician?
    1. verbal
    2. faxed
    3. in-person
    4. it depends on each State's pharmacy practice guidelines

114. What should you assume if the prescription does not indicate anything for refills?
    1. Assume PRN refills.
    2. Assume refills for 6 months.
    3. Assume no refills.
    4. Assume refills according to expiration date on the drug's bulk bottle label.

115. How are stock bottles of controlled substances distinguishable?
    1. The bottle contains a tamper-proof seal.
    2. Controlled substances may be purchased from specialty wholesalers who provide controlled substances information on the invoice.
    3. The label indicates a "C" and Roman numeral classification from II through V.
    4. The stock bottles may only be prepared in quantities of 90 to allow for a smaller sized bottle.

116. Which of the following is an example of a Class III controlled substance?
    1. acetaminophen with codeine
    2. oxycodone with acetaminophen
    3. oxazepam
    4. guaifenesin with codeine

117. In addition to the information typically found on a prescription order, the following information is required for all prescriptions for controlled substances
    1. patient's age, patient's gender
    2. patient's address, prescriber's DEA number
    3. patient's address, responsible payer
    4. prescriber's specialty, diagnosis

118. What should you consider when dealing with an elderly patient?
    1. hearing ability
    2. chronic conditions
    3. multiple drug management
    4. all of the above

119. What is verbal communication?
    1. messages transmitted through body motions
    2. messages delivered through touch
    3. messages transmitted the same as nonverbal communication
    4. messages using words

120. Which of the following best describes the attributes of an effective communicator?
    1. listens to the patient's concerns
    2. observes the patient for clues
    3. explains information on a level understood by the patient
    4. all of the above

121. What is effective communication?
    1. messages sent appropriately
    2. messages interpreted by the sender
    3. messages sent, received, and understood as the sender intended
    4. ongoing transmission of messages

122. Which of the following are references to common descriptions of 2 *general* types of patients who need pharmacy services?
    1. the elderly and women
    2. pediatrics and women of child-bearing age
    3. travelers and vacationers
    4. ambulatory and institutionalized

123. Which of the following best describes communications?
    1. directions provided on a prescription label
    2. an exchange of messages through verbal and nonverbal methods
    3. counseling offered by the pharmacist
    4. questions asked by the patient or caregiver

124. What is the importance of providing the patient with complete and understandable information about the directions and use of medications?
    1. It avoids potential misuse.
    2. It is required by State Boards of Pharmacy.
    3. It helps achieve the best therapeutic outcome.
    4. answers 1 and 3

125. Which of the following would be included in nonverbal communications?
    1. body movements
    2. facial expressions
    3. hand motions
    4. all of the above

126. Which of the following best describes proper telephone answering?
    1. Identify yourself and your organization.
    2. Get to the point immediately.
    3. Request the caller's phone number.
    4. Answer "Hello" and permit the caller to proceed.

127. What is appropriate telephone etiquette when you put a person on "hold"?
    1. Tell the person he or she must hold.
    2. Switch to on-line music.
    3. Assure the person that you will return.
    4. Let the person know you are busy.

Communications

128. Which of the following is not an important element associated with patient's rights?
    1. privacy
    2. confidentiality
    3. protection
    4. gender

129. Which of the following acronyms represent professional pharmacy associations?
    1. NAMES, ASPEN
    2. AMA, AHA
    3. APhA, NARD
    4. HCFA, HIDA

ANSWERS

I ASSISTING THE PHARMACIST IN SERVING PATIENTS

DRUGS

1. 3
2. 3
3. 1
4. 4
5. 2
6. 4
7. 1
8. 4
9. 3
10. 2
11. 1
12. 3
13. 1
14. 1
15. 4
16. 4
17. 2
18. 4
19. 1
20. 4

Answers

21. 1
22. 4
23. 4
24. 3
25. 2
26. 1
27. 2
28. 2
29. 4
30. 4
31. 1
32. 4
33. 4
34. 4
35. 2
36. 1
37. 2
38. 2
39. 2
40. 3
41. 2
42. 3
43. 2
44. 1
45. 2
46. 2
47. 4
48. 2
49. 3
50. 2
51. 3
52. 1
53. 4
54. 3
55. 3
56. 2
57. 1
58. 4
59. 1
60. 4
61. 4
62. 2
63. 3

Answers

64. 4
65. 3
66. 4
67. 2
68. 4
69. 2
70. 2
71. 4
72. 2
73. 4
74. 3
75. 2
76. 4
77. 1
78. 3
79. 3
80. 2
81. 4
82. 2
83. 3
84. 2
85. 4
86. 1
87. 3
88. 4
89. 4
90. 2
91. 4
92. 3
93. 4
94. 4
95. 4
96. 3
97. 3
98. 1
99. 2
100. 2
101. 4
102. 1
103. 4
104. 4
105. 4
106. 4

Answers

107. 3
108. 3
109. 4
110. 1
111. 2
112. 1
113. 1
114. 2
115. 2
116. 4
117. 3
118. 4
119. 4
120. 4
121. 4
122. 4
123. 2
124. 2
125. 1
126. 3
127. 3
128. 3
129. 1
130. 4
131. 4
132. 1
133. 2
134. 4
135. 2
136. 2
137. 3
138. 2
139. 4
140. 4
141. 3
142. 2
143. 2
144. 4
145. 2
146. 3
147. 1
148. 4
149. 4

150. 3
151. 2
152. 4
153. 1
154. 3
155. 1
156. 3
157. 3
158. 1
159. 3
160. 1
161. 3
162. 2
163. 3
164. 1
165. 4
166. 1
167. 1
168. 4
169. 3
170. 2
171. 4
172. 3
173. 1
174. 2
175. 4
176. 1
177. 4
178. 4
179. 3
180. 2
181. 2
182. 1
183. 1
184. 1
185. 4
186. 1
187. 2
188. 2
189. 4
190. 1
191. 4
192. 3

Answers

193. 2
194. 4
195. 1
196. 2
197. 1
198. 4
199. 3
200. 3
201. 1
202. 1
203. 3
204. 2
205. 3
206. 4
207. 1
208. 1
209. 1
210. 3
211. 4
212. 3
213. 2
214. 2
215. 4
216. 1
217. 4
218. 2
219. 4
220. 2
221. 2
222. 1
223. 2
224. 4
225. 4
226. 4
227. 4
228. 3
229. 1
230. 1
231. 4
232. 1
233. 1
234. 3
235. 4

236. 2
237. 4
238. 2
239. 2
240. 2
241. 3
242. 3
243. 1
244. 4
245. 2
246. 2
247. 2
248. 2
249. 3
250. 3
251. 1
252. 1
253. 1
254. 3
255. 3
256. 1
257. 1
258. 2
259. 3
260. 2
261. 4
262. 3
263. 3
264. 3
265. 3
266. 2
267. 4
268. 4
269. 3
270. 4
271. 4
272. 3
273. 2
274. 1
275. 4
276. 4
277. 4
278. 2

Answers

279. 4
280. 4
281. 2
282. 1
283. 1
284. 2
285. 2
286. 2
287. 1
288. 3
289. 3
290. 4
291. 4
292. 4
293. 3
294. 4
295. 4
296. 3
297. 3
298. 2
299. 1
300. 3
301. 2
302. 2
303. 2
304. 4
305. 1
306. 1
307. 2
308. 4
309. 1
310. 1
311. 2
312. 1
313. 4
314. 1
315. 2
316. 1
317. 3
318. 4
319. 2
320. 4
321. 2

322. 2
323. 2
324. 4
325. 4
326. 3
327. 2
328. 2
329. 4
330. 3
331. 4
332. 2
333. 4
334. 3
335. 2
336. 2
337. 3
338. 2
339. 1
340. 2
341. 1
342. 3
343. 3
344. 1
345. 3
346. 3
347. 4
348. 4
349. 3
350. 3
351. 4
352. 4
353. 4
354. 1
355. 4
356. 4
357. 4
358. 4
359. 2
360. 2
361. 3
362. 3
363. 4
364. 1

Answers

365. 2
366. 2
367. 4
368. 4
369. 2
370. 1
371. 3
372. 4
373. 1
374. 4
375. 1
376. 2
377. 1
378. 4
379. 2
380. 4
381. 1
382. 1
383. 4
384. 2
385. 2
386. 1
387. 2
388. 1
389. 2
390. 2
391. 4
392. 1
393. 3
394. 1
395. 4
396. 3
397. 3
398. 3
399. 1
400. 1
401. 4
402. 1
403. 4
404. 3
405. 1
406. 4
407. 2

408. 3
409. 2
410. 2
411. 3
412. 2
413. 3
414. 4
415. 4
416. 3
417. 3
418. 4
419. 4
420. 3
421. 3
422. 3
423. 4
424. 4
425. 3
426. 3
427. 2
428. 2
429. 3
430. 2
431. 1
432. 2
433. 3
434. 4
435. 3
436. 4
437. 2
438. 2
439. 3
440. 4
441. 1
442. 3
443. 4
444. 3
445. 2
446. 4
447. 3
448. 1
449. 2
450. 1

Answers

451. 4
452. 4
453. 2
454. 2
455. 2
456. 2
457. 1
458. 1
459. 2
460. 4
461. 4
462. 1
463. 3
464. 4
465. 4
466. 1
467. 4
468. 4
469. 1
470. 4
471. 4
472. 2
473. 3
474. 1
475. 4
476. 2
477. 1
478. 3
479. 4
480. 2
481. 3
482. 1
483. 2
484. 3
485. 4
486. 4
487. 1
488. 1
489. 1
490. 1
491. 3
492. 4
493. 4

494. 2
495. 4
496. 4
497. 2
498. 1
499. 4
500. 3
501. 3
502. 2
503. 2
504. 3
505. 1
506. 4
507. 2
508. 2
509. 4
510. 1
511. 3
512. 3
513. 4
514. 2
515. 1
516. 1
517. 4
518. 2
519. 1
520. 2
521. 4
522. 1
523. 2
524. 4
525. 2
526. 3
527. 3
528. 1
529. 2
530. 1
531. 1
532. 4
533. 2
534. 3
535. 4
536. 2

Answers

537. 2
538. 3
539. 2
540. 3
541. 2
542. 4
543. 1
544. 2
545. 1
546. 1
547. 2
548. 4
549. 4
550. 4
551. 2
552. 2
553. 3
554. 4
555. 1
556. 1
557. 1
558. 2
559. 4
560. 1
561. 2
562. 1
563. 4
564. 4
565. 1
566. 3
567. 2
568. 3
569. 1
570. 2
571. 2
572. 1
573. 3

I ASSISTING THE PHARMACIST IN SERVING PATIENTS

CALCULATIONS

1. 4
2. 2
3. 3
4. 2
5. 3
6. 3
7. 2
8. 4
9. 2
10. 3
11. 1
12. 2
13. 1
14. 3
15. 3
16. 4
17. 1
18. 3
19. 2
20. 1
21. 4
22. 4
23. 4
24. 3
25. 1
26. 1
27. 1
28. 2
29. 4
30. 2
31. 3
32. 1
33. 2
34. 2
35. 1
36. 1
37. 4
38. 2
39. 4
40. 1

Answers

41. 3
42. 4
43. 3
44. 3
45. 2
46. 3
47. 4
48. 4
49. 1
50. 3
51. 4
52. 1
53. 3
54. 1
55. 3
56. 2
57. 1
58. 3
59. 3
60. 4
61. 4
62. 1
63. 1
64. 2
65. 4
66. 1
67. 4
68. 3
69. 3
70. 4
71. 1
72. 3
73. 3
74. 4
75. 3
76. 2
77. 4
78. 3
79. 1
80. 2
81. 2
82. 1
83. 4

84. 4
85. 1
86. 1
87. 1
88. 3
89. 4
90. 3
91. 3
92. 2
93. 2
94. 3
95. 1
96. 2
97. 2
98. 2
99. 2
100. 3
101. 1
102. 1
103. 3
104. 1
105. 3
106. 3
107. 1
108. 4
109. 4
110. 2
111. 2
112. 2
113. 1
114. 3
115. 1
116. 1
117. 4
118. 2
119. 4
120. 4
121. 3
122. 3
123. 2
124. 3
125. 2
126. 3

Answers

127. 2
128. 4
129. 2
130. 4
131. 4
132. 2
133. 3
134. 1
135. 1
136. 4
137. 4
138. 1
139. 3
140. 2
141. 2
142. 4
143. 4
144. 2
145. 2
146. 1
147. 2
148. 2
149. 3
150. 1
151. 3
152. 2
153. 3
154. 2
155. 4
156. 4
157. 2
158. 2
159. 1
160. 3
161. 3
162. 3
163. 2
164. 1
165. 1
166. 4
167. 3
168. 2
169. 3

170. 1
171. 2
172. 4
173. 1
174. 3
175. 1
176. 4
177. 4
178. 4
179. 3
180. 3
181. 3
182. 4
183. 3
184. 2
185. 2
186. 3
187. 1
188. 3
189. 3
190. 2
191. 1
192. 2
193. 3
194. 4
195. 4
196. 1
197. 2
198. 3
199. 4
200. 1
201. 1
202. 2
203. 3
204. 4
205. 4
206. 3
207. 4
208. 2
209. 3
210. 4
211. 1
212. 2

Answers

213. 2
214. 1
215. 2
216. 3
217. 2
218. 2
219. 3
220. 4
221. 2
222. 3
223. 4
224. 3
225. 3
226. 1
227. 2
228. 4
229. 4
230. 3
231. 1
232. 2
233. 4
234. 2
235. 2
236. 4
237. 3
238. 4
239. 3
240. 4
241. 4
242. 2

I ASSISTING THE PHARMACIST IN SERVING PATIENTS

TERMINOLOGY

1. 3
2. 3
3. 4
4. 4
5. 1
6. 1
7. 2
8. 3

9. 2
10. 1
11. 4
12. 2
13. 2
14. 3
15. 1
16. 2
17. 4
18. 4
19. 1
20. 3
21. 1
22. 3
23. 2
24. 3
25. 2
26. 3
27. 1
28. 2
29. 3
30. 1
31. 4
32. 4
33. 1
34. 4
35. 3
36. 3
37. 1
38. 2
39. 3
40. 3
41. 1
42. 1
43. 4
44. 3
45. 2
46. 1
47. 4
48. 3
49. 3
50. 1
51. 1

Answers

52. 4
53. 1
54. 3
55. 4
56. 1
57. 4
58. 3
59. 2
60. 4
61. 1
62. 3
63. 3
64. 4
65. 3
66. 1
67. 2
68. 1
69. 1
70. 2
71. 4
72. 4
73. 2
74. 3
75. 4
76. 2
77. 3
78. 4
79. 1
80. 3
81. 3
82. 1
83. 2
84. 2
85. 3
86. 4
87. 4
88. 2
89. 1
90. 1
91. 1
92. 4
93. 4
94. 2

Answers

95. 3
96. 1
97. 3
98. 2
99. 2
100. 1
101. 3
102. 2
103. 2
104. 4
105. 3
106. 4
107. 3
108. 4
109. 2
110. 2
111. 1
112. 1
113. 4
114. 2
115. 4
116. 2
117. 3
118. 3
119. 4
120. 4
121. 2
122. 2
123. 2
124. 1
125. 2
126. 1
127. 4
128. 1
129. 2
130. 2
131. 4
132. 3
133. 4
134. 3
135. 1
136. 4
137. 2

Answers 195

138. 2
139. 4
140. 3
141. 4
142. 1
143. 2
144. 1
145. 2
146. 1
147. 1
148. 4
149. 4
150. 1
151. 4
152. 3
153. 3
154. 3
155. 4
156. 3
157. 3

I ASSISTING THE PHARMACIST IN SERVING PATIENTS

THE DISPENSING PROCESS

INFORMATION RESOURCES

1. 2
2. 4
3. 3
4. 3
5. 4
6. 1
7. 2
8. 3
9. 4
10. 3
11. 2
12. 3
13. 3
14. 1
15. 1
16. 3
17. 3

18. 4
19. 3
20. 2
21. 2
22. 1
23. 4
24. 2
25. 4
26. 3
27. 2
28. 1
29. 3
30. 1
31. 4
32. 4
33. 3
34. 1
35. 4
36. 2
37. 3
38. 1
39. 3
40. 4
41. 2
42. 4
43. 2
44. 3
45. 1
46. 4
47. 4
48. 3
49. 3
50. 1
51. 2
52. 1
53. 4
54. 4
55. 4
56. 2
57. 4
58. 4
59. 4
60. 4

Answers

61. 3
62. 3
63. 1
64. 1
65. 3
66. 3
67. 3
68. 2
69. 3
70. 3
71. 2
72. 4
73. 4
74. 3
75. 2
76. 3
77. 4
78. 3
79. 1
80. 4
81. 1
82. 1
83. 4
84. 4
85. 1
86. 3
87. 3
88. 3
89. 1
90. 1
91. 3
92. 1
93. 4
94. 2
95. 3
96. 2
97. 1
98. 3
99. 2
100. 4
101. 1
102. 1
103. 2

104. 2
105. 3
106. 3
107. 4
108. 2
109. 4
110. 2
111. 2
112. 4
113. 1
114. 3
115. 4
116. 2
117. 2
118. 3
119. 3
120. 4
121. 4
122. 1
123. 2
124. 2
125. 4
126. 3
127. 1
128. 3
129. 3
130. 2
131. 1
132. 3
133. 1
134. 2
135. 4
136. 4
137. 4
138. 2
139. 3
140. 2
141. 3
142. 2
143. 3
144. 1
145. 1
146. 1

Answers

147. 3
148. 4
149. 2
150. 4
151. 3
152. 1
153. 2
154. 2
155. 3
156. 3
157. 4
158. 1
159. 3
160. 3
161. 2
162. 4
163. 3
164. 2
165. 2
166. 1
167. 4
168. 1
169. 1
170. 4
171. 4
172. 4
173. 1
174. 4
175. 3
176. 3
177. 1
178. 2
179. 3
180. 1
181. 4
182. 3
183. 3
184. 4
185. 1
186. 2
187. 4
188. 2

189. 4
190. 1
191. 3
192. 3
193. 4
194. 2
195. 4
196. 2
197. 3
198. 3
199. 1
200. 3
201. 1
202. 4
203. 4
204. 3
205. 1
206. 4
207. 2
208. 3
209. 3
210. 2
211. 4
212. 3
213. 4
214. 3
215. 3
216. 2
217. 3
218. 1
219. 3
220. 2
221. 3
222. 1
223. 2
224. 2
225. 1
226. 1
227. 3
228. 3
229. 4

Answers

230. 4
231. 1
232. 2
233. 2

II MEDICATION DISTRIBUTION AND INVENTORY CONTROL SYSTEMS

PURCHASING

INVENTORY CONTROL

PREPARATION & DISTRIBUTION INVENTORY

1. 3
2. 3
3. 2
4. 1
5. 3
6. 2
7. 1
8. 4
9. 3
10. 2
11. 1
12. 1
13. 4
14. 2
15. 1
16. 1
17. 2
18. 2
19. 1
20. 2
21. 3
22. 2
23. 1
24. 3
25. 1
26. 3
27. 2
28. 2
29. 3
30. 3

31. 4
32. 2
33. 2
34. 3
35. 4
36. 2
37. 1
38. 3
39. 3
40. 1

III OPERATIONS MANAGEMENT

FACILITIES & MANAGEMENT

HUMAN RESOURCES

INFORMATION

LAW

COMMUNICATIONS

1. 2
2. 1
3. 4
4. 4
5. 1
6. 3
7. 2
8. 2
9. 1
10. 2
11. 4
12. 2
13. 2
14. 3
15. 3
16. 1
17. 1
18. 2
19. 2
20. 4

Answers

21. 2
22. 1
23. 4
24. 2
25. 1
26. 3
27. 4
28. 2
29. 3
30. 1
31. 4
32. 2
33. 3
34. 3
35. 4
36. 2
37. 3
38. 2
39. 2
40. 1
41. 1
42. 4
43. 4
44. 3
45. 2
46. 2
47. 4
48. 3
49. 4
50. 1
51. 2
52. 2
53. 3
54. 3
55. 4
56. 1
57. 3
58. 2
59. 1
60. 4
61. 3
62. 2
63. 1

64. 1
65. 1
66. 2
67. 3
68. 4
69. 4
70. 4
71. 2
72. 4
73. 4
74. 1
75. 3
76. 4
77. 1
78. 4
79. 1
80. 2
81. 4
82. 3
83. 2
84. 1
85. 2
86. 1
87. 1
88. 4
89. 4
90. 4
91. 2
92. 2
93. 2
94. 4
95. 1
96. 2
97. 4
98. 1
99. 4
100. 3
101. 2
102. 1
103. 3
104. 3
105. 4
106. 2